SPRINGER PROCEEDINGS IN PHYSICS 142

SPRINGER PROCEEDINGS IN PHYSICS

Please view available titles in *Springer Proceedings in Physics* on series homepage
http://www.springer.com/series/361/

Stefano Bellucci

Editor

Supersymmetric Gravity and Black Holes

Proceedings
of the INFN-Laboratori Nazionali
di Frascati School on the Attractor
Mechanism 2009

With 16 Figures

 Springer

Editor
Stefano Bellucci
Laboratori Nazionali di Frascati
Istituto Nazionale di Fisica Nucleare
Frascati RM
Italy

Springer Proceedings in Physics   ISSN 0930-8989
ISBN 978-3-642-31379-0   ISBN 978-3-642-31380-6 (eBook)
DOI 10.1007/978-3-642-31380-6
Springer Heidelberg New York Dordrecht London

Library of Congress Control Number: 2012953495

Printed on acid-free paper

Springer is part of Springer Science+Business Media (www.springer.com)

To Mario, a great physicist, a model of loyalty, integrity, coherence, and just the best Director I ever met.

Preface

This book is based upon lectures held from 29 June to 3 July 2009 at the INFN-Laboratori Nazionali di Frascati School on Attractor Mechanism, directed by Stefano Bellucci, with the participation of prestigious lecturers, including M. Cvetic, G. DallAgata, S. Ferrara, J.F. Morales, G. Moore, A. Sen, J. Simon, and M. Trigiante, as well as invited scientists of the caliber of M. Bianchi, C. Nappi, A. Sagnotti, and E. Witten. All lectures were given at a pedagogical, introductory level, a feature which reflects itself in the specific "flavor" of this volume, which also benefited much from extensive discussions and related reworking of the various contributions.

This is the fifth volume in a series of books on the general topics of supersymmetry, supergravity, black holes, and the attractor mechanism. Indeed, based on previous meetings, four volumes were already published:

Bellucci S. (2006). Supersymmetric Mechanics – Vol. 1: Supersymmetry, Non-commutativity and Matrix Models. (vol. 698, pp. 1–229). ISBN: 3-540-33313-4. (Springer, Berlin Heidelberg) Lecture Notes in Physics Vol. 698.

Bellucci S., S. Ferrara, A. Marrani. (2006). Supersymmetric Mechanics – Vol. 2: The Attractor Mechanism and Space Time Singularities. (vol. 701, pp. 1–242). ISBN: 978-3-540-34156-7. (Springer, Berlin Heidelberg) Lecture Notes in Physics Vol. 701.

Bellucci S. (2008). Supersymmetric Mechanics – Vol. 3: Attractors and Black Holes in Supersymmetric Gravity. (vol. 755, pp. 1–373). ISBN: 978-3-540-79522-3. (Springer, Berlin Heidelberg) Lecture Notes in Physics Vol. 755.

Bellucci S. (2010). The Attractor Mechanism. Proceedings of the INFN-Laboratori Nazionali di Frascati School 2007. ISSN 0930-8989, ISBN 978-3-642-10735-1, e-ISBN 978-3-642-10736-8. DOI 10.1007/978-3-642-10736-8. (Springer Heidelberg Dordrecht London New York) Proceedings in Physics Vol. 134.

I wish to thank all lecturers, invited scientists, and participants at the School for contributing to the success of the School, which prompted the realization of this volume. I wish to thank especially Mario Calvetti for giving vital support to the School and for personal trust and enduring encouragement. Lastly, but most importantly, my gratitude goes to my wife Gloria and our beloved daughters Costanza, Eleonora, Annalisa, Erica, and Maristella for love and inspiration, in want of which I would have never had the strength to complete this book.

Frascati, Italy   Stefano Bellucci

Contents

Chapter 1
Black Holes in Supergravity: Flow Equations and Duality

Gianguido Dall'Agata

1.1 Introduction

The analysis of black hole solutions and the study of their physics is an active and important branch of contemporary theoretical physics. In fact, not only black holes are an excellent theoretical laboratory for understanding some features of quantum gravity, but they can also be successfully used as a tool in applications to nuclear physics, condensed matter, algebraic geometry and atomic physics. For this reason, black holes are considered the "Hydrogen atom" of quantum gravity [67] or the "harmonic oscillator of the 21st century" [77].

The existence of black holes seems to be an unavoidable consequence of General Relativity (GR) and of its extensions (like supergravity). Classically, the horizon of black holes protects the physics in the outer region from what happens in the vicinity of singular field configurations that can arise in GR from smooth initial data. However, already at the semiclassical level, black holes emit particles with a thermal spectrum [7, 58]. A thermodynamic behaviour can also be associated to black holes from the laws governing their mechanics [79] and, in particular, one can associate to a black hole an entropy S proportional to the area A of its event horizon (measured in Planck units $l_P^2 = G\hbar/c^3$)

$$S = \frac{k_B}{l_P^2}\frac{A}{4}.$$

(1.1)

In most physical systems the thermodynamic entropy has a statistical interpretation in terms of counting microscopic configurations with the same macroscopic

G. Dall'Agata (✉)
Dipartimento di Fisica "Galileo Galilei", Università di Padova, Via Marzolo 8, 35131, Padova, Italy

INFN, Sefione di Padova, Via Marzolo 8, 35131, Padova, Italy
e-mail: dallagat@pd.infn.it

S. Bellucci (ed.), *Supersymmetric Gravity and Black Holes*, Springer Proceedings in Physics 142, DOI 10.1007/978-3-642-31380-6_1,
© Springer-Verlag Berlin Heidelberg 2013

properties, and in most cases this counting requires an understanding of the quantum degrees of freedom of the system. The identification of the degrees of freedom that the Bekenstein–Hawking entropy is counting is a long-standing puzzle that motivated much theoretical work of the last few years. String Theory, being a theory of quantum gravity, should be able to provide a microscopic description of black holes and hence justify Bekenstein–Hawking's formula. By now we have strong indications and many different and compelling examples where String Theory successfully accomplishes this goal, although often simplifying assumptions are made so that the configurations which are considered are not very realistic. In particular, black holes are non-perturbative objects and only for special classes of solutions (mainly supersymmetric) string theory at weak coupling can reproduce the correct answer[1] [33, 73, 78]. However, there is now a growing evidence that also for non-zero coupling we can identify candidate microstate geometries, whose quantization may eventually yield an entropy that has the same parametric dependence on the charges as that of supersymmetric black holes [5, 13, 65, 68].

In the last few years a lot of progress has been made in understanding the physics of *extremal non-supersymmetric solutions* and of their candidate microstates. The aim of these lectures is to provide an elementary and self-contained introduction to supergravity black holes, describing in detail the techniques that allow to construct full extremal solutions and to discuss their physical properties. We will especially focus on the peculiar role of scalar fields in supergravity models and on the flow equations driving them to the attractor point provided by the black hole horizon. We will also discuss the multicentre solutions and the role of duality transformations in establishing the classes of independent solutions.

1.2   Black Holes and Extremality

In this section we will review some general properties of black holes and discuss the concept of *extremality*, both in the context of geometrical and of thermodynamical properties of the solutions.

We will be interested in charged black hole configurations, so our starting point is the Einstein–Maxwell action in 4 dimensions, with Lagrangian density given by

$$e^{-1}\mathcal{L} = R - \frac{1}{4}\,F_{\mu\nu}F^{\mu\nu}. \tag{1.2}$$

For the sake of simplicity we will look for static, spherically symmetric and charged solutions. This means that the line element describing the metric should be of the form

---

[1]Recently there has been also a lot of progress in understanding the nature of the entropy for Kerr black holes and close to extremal examples of this sort can be realized in nature [20].

$$ds^2 = -e^{2U(r)}dt^2 + e^{-2U(r)}dr^2 + r^2 d\Omega^2, \tag{1.3}$$

where $d\Omega^2 = d\theta^2 + \sin^2\theta \, d\phi^2$ is the line element of a two-sphere and U is the warp factor, which depends only on the radial variable in order to respect spherical symmetry. For the same reason, the two-form associated to the Maxwell field $F_{\mu\nu}$ should be of the form

$$F = P \sin\theta \, d\theta \wedge d\phi + Q \, dt \wedge \frac{dr}{r^2}, \tag{1.4}$$

so that, by integrating over a sphere, one gets the electric and magnetic charge of the configuration:

$$\frac{1}{4\pi} \int_{S^2} F = P, \qquad \frac{1}{4\pi} \int_{S^2} \star F = Q. \tag{1.5}$$

By solving the equations of motion derived from (1.2) we obtain the following expression for the warp factor

$$e^{2U(r)} = 1 - \frac{2M}{r} + \frac{P^2 + Q^2}{r^2}, \tag{1.6}$$

which is the appropriate one for a Reissner–Nordström black hole and reduces to the one by Schwarzschild for $P = Q = 0$.

The solution above contains a singularity at $r = 0$, as one can see by computing the quadratic scalar constructed in terms of the Ricci tensor

$$R_{\mu\nu} R^{\mu\nu} = 4 \frac{(Q^2 + P^2)^2}{r^8} \xrightarrow{r \to 0} \infty \tag{1.7}$$

(For the special case $P = Q = 0$ we can still find a singularity in $R_{\mu\nu\rho\sigma} R^{\mu\nu\rho\sigma} = 48\frac{M^2}{r^6}$). However, the singularity is hidden by the horizons appearing at the zeros of the warp-factor function

$$e^{2U} = 0 \quad \Leftrightarrow \quad r_\pm = M \pm \sqrt{M^2 - (P^2 + Q^2)}. \tag{1.8}$$

The two solutions are real as long as $M^2 \geq P^2 + Q^2$, while the singularity becomes naked for smaller values of the mass. This means that, *for fixed charges, there is a minimum value of the mass for which the singularity is screened by the horizons.* At such value the warp factor has a double zero, the two horizons coincide and the semi-positive definite parameter

$$c = r_+ - r_- = \sqrt{M^2 - (P^2 + Q^2)}, \tag{1.9}$$

which we introduce for convenience, is vanishing. The corresponding black hole configuration is called *extremal* ($c = 0$ or $M = \sqrt{P^2 + Q^2}$). Note that in the

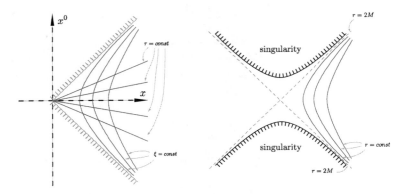

Fig. 1.1 Minkowski and Schwarzschild spacetimes in Rindler coordinates. The first diagram approximates the second close to the horizon

uncharged limit $c = M$, which is the extremality parameter for the Schwarzschild solution. This means that extremal Schwarzschild black holes are necessarily *small*, i.e. with vanishing horizon area at tree level.

Although the singularity is timelike (for charged solutions) and hence one can interpret it as the presence of a source, the existence of the horizons guarantees that the physics outside the horizon is not influenced by what happens inside, where one meets the singularity. This fact is easily seen by computing the time it takes for a light ray traveling radially to reach the horizon from infinity, as measured by an observer sitting far from the black hole. By taking $ds = 0$ for constant θ and ϕ one gets that

$$\sqrt{g_{tt}}\, dt = \sqrt{g_{rr}}\, dr, \tag{1.10}$$

so that the time it takes for a light ray to travel radially between two points at distance r_1 and r_2 from the singularity is proportional to the distance measured with a weight given by the inverse of the warp factor

$$t_{12} = \int_{r_1}^{r_2} \sqrt{\frac{g_{rr}}{g_{tt}}}\, d\tilde{r} = \int_{r_1}^{r_2} e^{-2U(\tilde{r})}\, d\tilde{r}. \tag{1.11}$$

This expression goes to infinity when $r_1 \to r_+$ and therefore a signal from the horizon takes an infinite time to reach a far distant observer.

The physics close to the horizon can be better understood by considering the expansion of the solution obtained above for r close to r_+ (Fig. 1.1). The only non-trivial function in the metric is given by the warp factor, which approaches

$$e^{2U} = \frac{(r - r_+)(r - r_-)}{r^2} \xrightarrow{r \to r_+} \frac{r_+ - r_-}{r_+^2}\, \rho, \tag{1.12}$$

where we introduced a new coordinate ρ measuring the distance from the outer horizon: $\rho = r - r_+$. The resulting near horizon geometry is

$$ds^2 \to -\frac{r_+ - r_-}{r_+^2}\rho \, dt^2 + \frac{r_+^2}{r_+ - r_-}\frac{d\rho^2}{\rho} + r_+^2 \, d\Omega^2, \tag{1.13}$$

which can be interpreted as the product of a 2-dimensional Rindler spacetime with a two-sphere of radius r_+. We can actually make this result explicit by performing another change of coordinates $(t, \rho) \mapsto (\tau, \xi)$ as follows

$$\rho = e^{2\alpha\xi}, \qquad t = \frac{1}{4\alpha^2}\tau, \qquad \alpha = \frac{\sqrt{r_+ - r_-}}{2r_+}. \tag{1.14}$$

This leads to a near-horizon metric described by

$$ds_{NH}^2 = e^{2\alpha\xi}\left(-d\tau^2 + d\xi^2\right) + r_+^2 \, d\Omega^2. \tag{1.15}$$

The geometry of the non-compact part is 2-dimensional Minkowski spacetime as seen by an observer that is uniformly accelerated with acceleration $\alpha = \sqrt{\alpha_\mu \alpha^\mu}$. In fact the change of coordinates from the standard ones to Rindler's is dictated by the trajectory of an accelerated observer

$$x(x^0) = \frac{1}{\alpha}\sqrt{1 + \alpha^2(x^0)^2}, \tag{1.16}$$

and τ denotes the proper time

$$x^0(\tau) = \frac{1}{\alpha}\sinh(\alpha\tau). \tag{1.17}$$

Our derivation explains this acceleration as the effect of gravitation and one can actually show that α coincides with the surface gravity of the black hole. In fact surface gravity is given in terms of the derivative of the null Killing vector generating the horizon surface, computed at the surface [79]

$$\alpha^2 = \left[-\frac{1}{2}\nabla_\mu\xi_\nu \nabla^\mu\xi^\nu\right]_{r=r_+} \tag{1.18}$$

and the two expressions coincide.

1.2.1 Thermodynamics

Hawking and Unruh showed that an accelerated observer following the trajectory described in (1.16) sees a thermal spectrum with temperature proportional to the acceleration:

$$T = \frac{\alpha}{2\pi}. \tag{1.19}$$

A simple heuristic argument to understand this result follows by considering the near-horizon geometry rotated to Euclidean signature. This is the same procedure that is used to describe quantum field theories with temperature. In fact, in quantum mechanics, for a system with Hamiltonian H, the thermal partition function is

$$Z = \text{Tr} \, e^{-\beta H}, \tag{1.20}$$

where β is the inverse temperature and Z is related to the time evolution operator $e^{-i\tau H}$ by a Euclidean analytic continuation. From the geometric point of view, by defining $\tilde{\tau} = i\tau$ and $\tilde{\xi} = e^{\alpha \xi}/\alpha$, the resulting euclidean geometry is

$$ds^2 = d\tilde{\xi}^2 + \alpha^2 \, \tilde{\xi}^2 \, d\tilde{\tau}^2, \tag{1.21}$$

which has a conical singularity at the origin unless $\tau \sim \tau + i\beta$, where

$$\beta = \frac{1}{T} = \frac{2\pi}{\alpha} = \frac{4\pi r_+^2}{r_+ - r_-} \tag{1.22}$$

and this gives an expression of the temperature in terms of the geometric quantities defining the black hole horizons.

Having now a thermodynamic system for which we defined the energy (given by the mass of the black hole M) and a temperature T, it is natural to define a (Bekenstein–Hawking) entropy S_{BH}, such that, for fixed charges, one fulfills the thermodynamic relation

$$\frac{dS_{BH}}{dM} = \frac{1}{T}. \tag{1.23}$$

In the case at hand, namely the Reissner–Nordström black hole, integration of the previous equation leads to

$$S_{BH} = \pi \, r_+^2 = \pi \left[M + \sqrt{M^2 - (P^2 + Q^2)} \right]^2. \tag{1.24}$$

The dependence of the entropy on the mass and charge of the black hole is summarized by the geometric quantity r_+, the horizon radius, which can be translated to the horizon area, leading to the famous relation

$$S_{BH} = \frac{A}{4}, \tag{1.25}$$

which is also valid for other configurations at the two-derivatives level. This is a remarkable relation between the thermodynamic properties of a black hole on the one hand and its geometric properties on the other and it is a cornerstone for our understanding of any theory of quantum gravity. In fact, if we believe that S_{BH} has the meaning of a real entropy, although such a quantity is usually defined in terms of global properties of the system, it contains non trivial information about the microscopic structure of the theory via Boltzmann's relation

$$S = \log \Omega, \tag{1.26}$$

where Ω is the total number of microstates of the system for a given energy and fixed total charges. In detail, the entropy contains information about the total number of microscopic degrees of freedom of the system and in our case a microscopic theory of gravity should explain the black hole entropy in terms of the quantum numbers defining the solution:

$$S_{BH} = \log \Omega(M, Q, P). \tag{1.27}$$

Explaining this formula is actually one of the biggest problems in theoretical physics. Given (1.25) relating the entropy of a black hole to its horizon area, we can actually see that the typical number of microstates forming a black hole is humongous. For instance, the black hole at the centre of our galaxy (Sgr A*) is estimated to have a radius of about $r_+ \sim 7 \cdot 10^9$ km [52], leading to an estimate of $S_{BH} \sim 10^{100}$, and this is just the logarithm of the number of states defining the black hole! If, on the other hand, we think about a generic black hole solution in GR, we know that the no-hair theorem tells us that a black hole is completely specified by its mass and charges. This would mean that, for fixed mass and charges, there is a unique classical state, leading to an expectation of $S = 0$.

It is actually interesting that the laws of black hole mechanics can be put in a one to one relation with the laws of thermodynamics [79]:

- Zeroth law: the temperature of a black hole $T = \alpha/2\pi$ is uniform at the horizon;
- First law: for quasi static changes the energy (mass) of a black hole changes as

$$dM = TdS + \psi dQ + \chi dP + \Omega dJ - g_{ij} \Sigma^i d\phi^j, \tag{1.28}$$

  where the entropy is identified with the area of the horizon as in (1.25), Q and P are the electric and magnetic charges, J is the angular momentum, ψ, χ and Ω are the associated chemical potentials (namely the electric and magnetic potentials at the horizon and the angular velocity, assumed constant for stationary solutions), ϕ^i the values of the scalar fields defining the solution and Σ^i the scalar charges;
- Second law: the horizon area always increases in time $\Delta A \geq 0$. A consequence of this last law is that coalescence processes are possible, while generically no splitting processes are allowed. For instance, two Schwarzschild black holes with masses M_1 and M_2 can form a bigger black hole with mass $M_1 + M_2$ because their horizon area is proportional to the square of the corresponding masses and therefore $(M_1 + M_2)^2 \geq M_1^2 + M_2^2$. The inverse process is forbidden by the same argument.

Coming back to our example, the Reissner–Nordström black hole, we can see that the temperature T, the entropy S_{BH} and the extremality parameter c are all defined in terms of the characteristic geometric quantities of the solution, namely the radii of the two horizons. This implies that, by comparing (1.24), (1.22) and (1.9), we can express the extremality parameter in terms of the temperature and entropy as

$$T = \frac{\alpha}{2\pi} = \frac{c}{2S} \quad \Rightarrow \quad c = 2\,S\,T. \tag{1.29}$$

Recalling that an extremal configuration is such when the two horizons coincide, i.e. $c = 0$, and that the entropy is non-vanishing whenever there is a non-trivial horizon $S = \pi\, r_+^2$, we can see that extremality implies vanishing temperature:

$$\text{Extremality} \quad \Leftrightarrow \quad c = 2\,ST = 0 \quad \Rightarrow \quad T = 0. \tag{1.30}$$

Extremal black holes are therefore thermodynamically stable. They do not radiate. We will come back to an explanation of this fact momentarily.

The special properties of this kind of black holes is reflected also in the near horizon metric, which now is not given by (1.15) anymore. Since the warp factor has a double zero, its behaviour close to the horizon is approximated by a quadratic function of ρ rather than linear as in (1.12)

$$e^{2U} = \frac{(r_+ - r_-)^2}{r^2} \quad \rightarrow \quad \frac{\rho^2}{r_+^2}. \tag{1.31}$$

The near-horizon metric changes accordingly and, by introducing

$$z = -\frac{M^2}{\rho}, \tag{1.32}$$

we can see that it is given by the product of a 2-dimensional Anti-de Sitter spacetime and a 2-dimensional sphere, both with radius $M = \sqrt{P^2 + Q^2}$:

$$ds_{NH}^2 = M^2 \left(\frac{-dt^2 + dz^2}{z^2} \right) + M^2 d\Omega^2. \tag{1.33}$$

Remember that, using these coordinates, the horizon sits at $z \rightarrow -\infty$. It is interesting to note that this geometry is conformally flat (extremal Reissner–Nordström solutions are also supersymmetric).

Before proceeding to a more detailed analysis of the differences between extremal and non-extremal black holes, let us pause for a second to make some comments. From the above discussion we can see that black holes are rather special thermodynamic systems, because they do not satisfy Nernst law, which states that the entropy should vanish (or arrive at a "universal constant" value) as the temperature approaches zero. The analog of this law fails in black hole mechanics, because extremal black holes have vanishing temperature, but non-vanishing entropy, $S = \pi \sqrt{P^2 + Q^2}$ in the previous example. However, there is good reason to believe that "Nernst theorem" should not be viewed as a fundamental law of thermodynamics but rather as a property of the density of states near the ground state in the thermodynamic limit, which happens to be valid for commonly studied materials. Indeed, examples can be given of ordinary quantum systems that violate the Nernst form of the third law in a manner very similar to the violations of the analog of this law that occur for black holes [80].

Another interesting observation follows from rewriting the metric ansatz in an isotropic form:

$$ds^2 = -H^{-2}(\boldsymbol{x})\,dt^2 + H^2(\boldsymbol{x})d\boldsymbol{x}_3^2. \tag{1.34}$$

Once the metric is written in this fashion, the equations of motion for the warp factor can be expressed as

$$\triangle_3 H = 0 \tag{1.35}$$

and therefore can be solved by generic harmonic functions, which may have more than one centre:

$$H = 1 + \sum_i \frac{m_i}{|\boldsymbol{x} - \boldsymbol{x}_i|}, \qquad m_i = \sqrt{p_i^2 + q_i^2}, \tag{1.36}$$

where $\boldsymbol{x}_i$ denotes the position of the i-th centre. This solution is allowed by the fact that gravitational attraction equals electromagnetic repulsion for each centre and hence leads to a condition of static neutral equilibrium. The additive nature of the solution is related to the BPS nature of force-free objects.

Finally, the fact that the near-horizon geometry approaches the product of an Anti-de Sitter spacetime and a sphere is actually a universal behaviour of extremal p-branes in D dimensions, whose near horizon geometry is given by $AdS_{p+2} \times S^{D-p-2}$. Black holes in four spacetime dimensions are a simple instance where $p = 0$ and $D = 4$, but one could also think of different examples like black holes and black strings ($p = 0, 1$) in $D = 5$, dyonic black strings ($p = 1$) in $D = 6$ and D3-branes in IIB string theory ($p = 3, D = 10$).

1.2.2 Extremal Versus Non-extremal Solutions

We can now go back to the concept of extremality to discuss an important difference between extremal and non-extremal black holes. A general ansatz for the metric that satisfies the requirements of describing spherically symmetric, static, asymptotically flat black holes and which encompasses both the extremal and non-extremal solutions is the following:

$$ds^2 = -e^{2U}\,dt^2 + e^{-2U}\left[\frac{c^4}{\sinh^4(cz)}dz^2 + \frac{c^2}{\sinh^2(cz)}d\Omega^2\right]. \tag{1.37}$$

The extremality parameter c was explicitly inserted and the extremal case is recovered by sending $c \to 0$, so that the metric simplifies to

$$ds_{c=0}^2 = -e^{2U}\,dt^2 + e^{-2U}\left[\frac{dz^2}{z^4} + \frac{1}{z^2}d\Omega^2\right], \tag{1.38}$$

where one can rewrite the factor in brackets using isotropic coordinates as a plain $\mathbb{R}^3$:

$$ds_{c=0}^2 = -e^{2U}\,dt^2 + e^{-2U}\,d\boldsymbol{x}^2. \tag{1.39}$$

Fig. 1.2 Schematic
representation of
non-extremal and extremal
black hole throats using
proper-distance coordinates

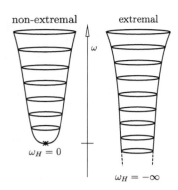

By applying this ansatz to the Reissner–Nordström case analyzed before, it is easy
to realize that the horizon sits at $z \to -\infty$.

However, proper distance from the horizon has to be computed by using
appropriate coordinates. In the non-extremal case, assuming that the horizon area
is finite, one gets that the factor in front of the angular variables should remain finite
as $z \to -\infty$ (Fig. 1.2).

This means that

$$e^{-2U} \frac{c^2}{\sinh^2(cz)} \xrightarrow{z \to -\infty} \frac{A}{4\pi} = r_H^2, \tag{1.40}$$

where r_H is the radius of the horizon. A proper radial coordinate ω can then be
introduced by considering the g_{zz} component of the metric in the same limit:

$$e^{-2U} \frac{c^4}{\sinh^4(cz)} dz^2 \longrightarrow \frac{A}{4\pi} 4c^2 e^{2cz} dz^2 \equiv r_H^2 d\omega^2. \tag{1.41}$$

Distances should then be measured by $\omega = 2\,e^{cz}$ in units of r_H and the black hole
horizon sits at $\omega_H = 0$, at finite proper distance from an arbitrary observer

$$L = \int_{\omega_H}^{\omega_0} r_H d\omega = r_H \omega_0 < \infty. \tag{1.42}$$

On the other hand, having finite area in the extremal case means

$$\frac{e^{-2U}}{z^2} \longrightarrow \frac{A}{4\pi} = r_H^2. \tag{1.43}$$

This implies that a new proper radial coordinate can be introduced by identifying

$$e^{-2U} \frac{dz^2}{z^4} \longrightarrow \frac{A}{4\pi} \frac{dz^2}{z^2} = r_H^2 \, d\omega^2, \tag{1.44}$$

which means
$$\omega = -\log(-z).\tag{1.45}$$
The horizon is now at $\omega_H = -\infty$ at infinite proper distance from any observer
$$L = \int_{\omega_H}^{\omega_0} r_H d\omega = +\infty.\tag{1.46}$$

As we will see in a moment, this difference has a crucial impact on the behaviour of scalar fields in this scenario and implies the existence of an attractor mechanism for extremal black hole configurations. Moreover, the fact that the horizon is at infinite proper distance from any observer justifies also the fact that extremal black holes are thermodynamically stable. Any radiation emitted by such black hole would be infinitely red-shifted before reaching any observer outside the horizon.

1.3 Attractors

1.3.1 Black Holes and Scalar Fields

When dealing with supergravity theories, as with many other effective theories of fundamental interactions, gravity needs to be coupled to scalar fields, possibly parameterizing a scalar σ-model and affecting also the couplings of the vector fields (which we consider abelian for the sake of simplicity). A generic Lagrangian describing the bosonic degrees of freedom of such theories will have the form

$$e^{-1}\mathcal{L} = R - \frac{1}{2}g_{ij}(\phi)\,\partial_\mu\phi^i\partial^\mu\phi^j + \frac{1}{4}\mathcal{I}_{\Lambda\Sigma}(\phi)F_{\mu\nu}^\Lambda F^{\Sigma\,\mu\nu} + \frac{1}{4}\mathcal{R}_{\Lambda\Sigma}(\phi)\frac{\epsilon^{\mu\nu\rho\sigma}}{2\sqrt{-g}}F_{\mu\nu}^\Lambda F_{\rho\sigma}^\Sigma,\tag{1.47}$$

where $g_{ij}(\phi)$ is the metric of the scalar σ-model, $\mathcal{I}$ is definite negative and describes the gauge kinetic couplings, $\mathcal{R}$ is the generalization of the θ-angle terms in the presence of many scalar and vector fields and we assume that there is no scalar potential for the time being. We are still interested in finding single centre, static, spherically symmetric, charged and asymptotically flat black hole solutions and therefore we keep the metric ansatz (1.37) and the request that the integral of the vector field strengths and their duals on a sphere at infinity gives the electric and magnetic charges of the solution:

$$\frac{1}{4\pi}\int_{S^2}F^\Lambda = p^\Lambda,\qquad\qquad \frac{1}{4\pi}\int_{S^2}G_\Lambda = q_\Lambda.\tag{1.48}$$

Now, however, we need to introduce a new definition of the dual field strengths G_Λ, because the values of the gauge couplings and the charges will be affected by the values of the scalar fields appearing in the functions $\mathcal{R}$ and $\mathcal{I}$.

In the setup considered in the previous section, the magnetic and electric charges are associated to the two-forms appearing in the Bianchi identities and in the equations of motion of the electromagnetic theory, respectively. These two-forms are also related between them by the known electromagnetic duality $F \leftrightarrow \star F$. In a general setup, like the one considered here, electric–magnetic duality can be extended to a new group of duality transformations that leaves invariant Bianchi identities and equations of motion [48]. If we focus on the part of the Lagrangian involving the gauge field-strengths

$$S_{EM} = \int \left[\mathcal{I}_{\Lambda\Sigma} F^{\Lambda} \wedge \star F^{\Sigma} - \mathcal{R}_{\Lambda\Sigma} F^{\Lambda} \wedge F^{\Sigma} \right], \tag{1.49}$$

we can deduce that the Bianchi identities and equations of motion form a set, from which we can define the dual field strengths G_{Λ}:

$$\begin{cases} dF^{\Lambda} = 0, \\ dG_{\Lambda} = d \left(\mathcal{R}_{\Lambda\Sigma} F^{\Sigma} - \mathcal{I}_{\Lambda\Sigma} \star F^{\Sigma} \right) = 0. \end{cases} \tag{1.50}$$

It is obvious, that for any constant matrix S we can rotate the original field strengths F^{Λ} and the dual ones G_{Λ} between them, leaving the full set of Bianchi identities and equations of motion invariant:

$$\begin{pmatrix} F \\ G \end{pmatrix} \rightarrow \begin{pmatrix} F' \\ G' \end{pmatrix} = S \begin{pmatrix} F \\ G \end{pmatrix}. \tag{1.51}$$

However, the requirement that also the definition of the dual field-strengths $G_{\Lambda} \equiv -\frac{\delta \mathcal{L}}{\delta F^{\Lambda}}$ remains invariant constrains the duality transformation to be part of the symplectic group $S \in \mathrm{Sp}(2n_V, \mathbb{R})$, where n_V is the total number of abelian vector fields in the theory and the symmetry transformations are continuous at the classical level. Moreover additional matter couplings, like the ones considered in (1.47), may reduce it to $G \subset \mathrm{Sp}(2n_V, \mathbb{R})$. G is called the U-duality group of the theory. An important result of [48] is that the stress energy tensor and hence the Einstein equations of motion following from rather general interactions between the various fields are invariant under such transformations. This means that by using the duality group we can map solutions of the Bianchi identities and of the equations of motion to new solutions of the same set of equations, leaving the metric untouched. In particular, we can map charged black hole solutions with different charges and scalar fields between them without changing the metric and hence a crucial property like the area of the horizon.

Before proceeding, let us note that by performing such duality transformations the Lagrangian does not necessarily remain invariant. U-duality transformations are symmetry transformations by which the equations of motion and the Bianchi identities are mixed among themselves linearly and this may require changes in the Lagrangian originating them.

Since (F^Λ, G_Λ) form a symplectic vector of closed two-forms, we could introduce explicitly the corresponding potentials $(A^\Lambda_\mu, A_{\mu\Lambda})$, though obviously not both at the same time. Given the request (1.48), the vector potentials should have a restricted form such that the integrals provide the correct electric and magnetic charges. In particular, solving $dF^\Lambda = 0$ and respecting the request of finding solutions with spherical symmetry, we can introduce

$$A^\Lambda = \chi^\Lambda(r)dt - p^\Lambda \cos\theta\, d\phi, \tag{1.52}$$

where χ^Λ are the electric potentials, so that $F^\Lambda = dA^\Lambda$. In the same fashion we can also introduce the dual potentials

$$A_\Lambda = \psi_\Lambda(r)dt - q_\Lambda \cos\theta\, d\phi, \tag{1.53}$$

where ψ_Λ are the magnetic potentials, so that $G_\Lambda = dA_\Lambda$. In the original action (1.47) only the first appears. Actually, we can see that χ^Λ appears in the action only under derivatives and therefore we can integrate it out. In fact, from the χ^Λ equations of motion one gets

$$\chi^{\Lambda\prime} = e^{2U} \mathcal{I}^{-1\,\Lambda\Sigma}\left(q_\Sigma - \mathcal{R}_{\Sigma\Gamma}\, p^\Gamma\right), \tag{1.54}$$

which is also the correct relation needed to fulfill the duality relation by which the definition of G_Λ follows from the one of F^Λ.

A simple strategy to find black hole solutions in this framework is to use the fact that the problem is spherically symmetric, so that all relevant quantities depend only on the radial variable, $\phi^i = \phi^i(r)$, $U = U(r)$, etc., and reduce the system to 1 dimension. By using (1.54) and by integrating out formally θ, ϕ and t one gets the effective 1-dimensional action

$$L_{1d} = (U')^2 + \frac{1}{2} g_{ij}\phi^{i\prime}\phi^{j\prime} + e^{2U} V_{BH} - c^2, \tag{1.55}$$

where

$$V_{BH} = -\frac{1}{2}Q^T \mathcal{M} Q, \tag{1.56}$$

with

$$\mathcal{M} = \begin{pmatrix} I + RI^{-1}R & -RI^{-1} \\ -I^{-1}R & I^{-1} \end{pmatrix} \tag{1.57}$$

and

$$Q = \begin{pmatrix} p^\Lambda \\ q_\Lambda \end{pmatrix}. \tag{1.58}$$

The kinetic term for the warp factor and the overall constant c^2 come from the reduction of the Einstein kinetic term. The black hole potential V_{BH} comes from the reduction of the kinetic term and θ-angle terms of the vector fields, after dualization of the electric potential using (1.54). The resulting problem is a 1-dimensional mechanical system of $n + 1$ scalars in the presence of a potential V_{BH} depending on a number of parameters equal to the total number of non-vanishing electric and magnetic charges.

Since we have made an ansatz on the metric, we should take into account the possibility that the equations of motion of this reduced system do not solve also the equations of motion of the original one, because we are essentially looking at constrained variations of the original system. This is actually the case and it means that in order to obtain solutions from the equations of motion coming from this Lagrangian that are equivalent to the original ones, we need to supplement the 1-dimensional Lagrangian (1.55) with the constraint

$$(U')^2 + \frac{1}{2} g_{ij} \phi^{i\prime} \phi^{j\prime} = e^{2U} V_{BH} + c^2. \tag{1.59}$$

For completeness, we provide here the equations of motion:

$$U'' = e^{2U} V_{BH}, \tag{1.60}$$

$$\phi^{i\prime\prime} + \Gamma_{jk}{}^{i} \phi^{j\prime} \phi^{k\prime} = e^{2U} g^{ij} \partial_j V_{BH}. \tag{1.61}$$

1.3.2 General Features of the Attractor Mechanism

We have seen that black holes in generic supergravity theories will depend on scalar fields. However, extremal black holes have the special property that the horizon quantities loose all the information about them. This is true independently of the fact that the solution preserves any supersymmetry or not. The horizon is in fact an attractor point [37, 38, 40, 76]: scalar fields, independently of their value at spatial infinity, flow to a fixed point given in terms of the charges of the solution at the horizon. Recalling that the entropy of black holes is given by the area of the horizon, this attractive behaviour for the scalar fields implies that for extremal black holes the entropy is a topological quantity, given in terms of quantized charges and therefore it does not depend on continuous parameters, which is a very appealing feature in order to have the chance to provide a microscopic explanation for the resulting number.

The main reason at the base of the attractor mechanism is the fact that for extremal black holes the horizon is at an infinite proper distance from any observer [41]. This means that while moving along the infinite throat leading to the horizon, scalar fields lose memory of the initial conditions. This is an obvious outcome of the request of having regular solutions. In fact, regularity of the scalar fields at the horizon implies that their derivative should vanish while approaching the horizon

$$\phi^{i\prime} \overset{z \to -\infty}{\longrightarrow} 0. \tag{1.62}$$

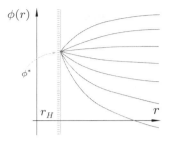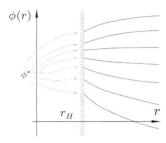

Fig. 1.3 Generic behaviour of a scalar field in the case of an extremal black hole (*left*) and of a non-extremal one (*right*). In the first case the scalar field ϕ runs towards the same value at the horizon ϕ^*, no matter what was its value at infinity. In the second case the scalar stops at different points in moduli space depending on its asymptotic value

On the other hand non-extremal black holes have $\omega_H = 0$ and hence scalar fields do not have time to blow up even for a non-trivial, but finite, first derivative along the radial direction.

Extremal black holes are therefore described by trajectories in the moduli space with a fixed point reached when the proper radial parameter $\omega \to -\infty$. The fixed point is an attractor of the system (Fig. 1.3). Since infinitely far away from the black hole ($\omega \to +\infty$) the geometry approaches that of 4-dimensional Minkowski spacetime and at the horizon ($\omega \to -\infty$) it approaches the product AdS$_2 \times S^2$, we can see that extremal black holes can also be thought of as solitons interpolating between two different vacua of the theory. We will come back to this picture later on to justify the description of such solutions in terms of first-order differential equations.

From the equations of motion of the scalar fields (1.61), fixed scalars at the horizon imply that the moduli reached a critical point of the black hole potential:

$$\partial_i V_{BH}(\phi^{i*}, q, p) = 0. \tag{1.63}$$

This is actually an intrinsic characterization of the horizon for extremal black holes. Extremization of the scalar potential, means that the scalar fields at the horizon take the value ϕ^{i*} such that the minimization condition (1.63) is satisfied. Since the only parameters appearing in the minimization condition are the black hole charges, the resulting attractor values of the moduli fields are also going to be given in terms of the same charges

$$\phi^{i*} = \phi^{i*}(q, p). \tag{1.64}$$

In turn, this implies that at the horizon the value of the scalar potential does not depend anymore on the values of the scalars at infinity, but only on the charges:

$$V_{BH}^* = V_{BH}(\phi^{i*}(q, p), q, p). \tag{1.65}$$

Fig. 1.4 Examples of black
hole potentials for various
values of the quartic
invariants

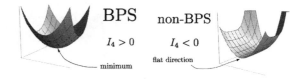

At this point we can also solve the equation for the warp factor (1.60) close to the
attractor point, which gives that

$$U \to -\log\left(\sqrt{V_{BH}^*} z\right).$$ (1.66)

This implies that the metric approaches that of $AdS_2 \times S^2$ as expected, with a
characteristic horizon radius given by $r_H = \sqrt{V_{BH}^*}$. This in turn implies that the
entropy of extremal black holes can be expressed in terms of the value of the black
hole potential at the critical point:

$$S_{BH} = \frac{A}{4} = \pi V_{BH}^*(q, p).$$ (1.67)

Since the black hole potential depends only on the quantized charges, because of
the attractor mechanism, also the entropy of extremal black holes will depend on
the same quantized parameters and all the possible dependence on the value of the
moduli fields (which still characterize the full solution) is lost.

The contrast becomes even more clear if we compare (1.67) with the corresponding expression for non-extremal solutions, where the area formula is valid for a
radius of the horizon sitting at the larger value between

$$r_\pm = M \pm \sqrt{M^2 - V_{BH}(\phi_\infty, p, q) + \frac{1}{2} g_{ij}(\phi_\infty) \Sigma^i \Sigma^j}.$$ (1.68)

Not only this expression depends on the value of the scalar fields at infinity, but also
on the scalar charges, which vanish only for solutions where the scalars remain
constant [53].

An interesting outcome of this analysis is that extremal black hole solutions
are completely specified by the black hole potential. In particular different kind of
attractors will be characterized by different types of potentials. Still, while V_{BH}
depends on the theory under investigation, the general features of the attractor
mechanism are universal.

As an example, consider the most constrained supergravity theory in 4 dimensions: maximally supersymmetric ($N = 8$) supergravity. This theory has a fixed
matter content, which is all contained in the gravity multiplet. Among other fields,
the gravity multiplet contains 28 vector fields, leading to 56 charges, and 70 scalar
fields parameterizing the scalar manifold $E_{7(7)}/SU(8)$.

Fig. 1.5 Runaway behaviour
for small black holes

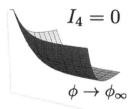

$$I_4 = 0$$

$$\phi \to \phi_\infty$$

The black hole potential depends on the detailed choice of the 28 electric and 28 magnetic charges, but, given the restrictive form of the scalar manifold and of the invariances of the theory, one can distinguish three main classes of solutions. These are related to the value of a special $E_{7(7)}$ invariant, which is quartic in the charges $I_4(p, q)$ [36, 63]. Whenever $I_4 > 0$ the scalar potential has a minimum and the corresponding black hole solutions preserve some supersymmetry (Fig. 1.4). If $I_4 < 0$ the solutions are non-supersymmetric. Finally, in the special instance where the quartic invariant vanishes, the warp factor at the horizon vanishes. This implies that the corresponding classical geometry is singular and various orbits can be further distinguished by the values of derivatives of I_4. However, higher-order corrections in the curvature terms modify the equations of motion in a way such that a horizon is developed, with a characteristic radius of the order of the typical scale of the correction terms. For this reason the corresponding black holes are called small black holes (Fig. 1.5).

In generic $N = 2$ theories supersymmetric configurations are always minima, while non-BPS ones have flat directions at the attractor point the potential (actually these flat directions are generically given by expectation values of scalar fields that do not appear in the scalar potential at all) [39]. For $N > 2$ also supersymmetric attractors may have a non-trivial moduli space.

One important lesson that can be learned from this analysis is that while supersymmetry always implies extremality, the opposite is not true. In fact the supersymmetry condition is achieved when the mass of the BPS object equals a certain value defined by its charges, however, for a given charge configuration the BPS bound may never be reached and hence for such configurations the object with the minimal mass will still be extremal, though non supersymmetric.

1.4   Glancing Through Special Kähler Geometry

As discussed in the previous section, the black hole potential containing the necessary information to describe extremal black holes depends on the detail of the model under investigation. General features of these solutions can anyway be obtained independently on such details. Although in the following we will try to give some general arguments about the properties of single and multi-centre extremal

black holes in supergravity theories, it is better to fix a specific framework, so that
we can provide explicit examples along with the general arguments. For this reason
we now provide a brief *intermezzo* with some elementary facts about Special Kähler
geometry, which is the geometric structure underlying the vector multiplet scalar
σ-model in $\mathcal{N} = 2$ theories in 4 dimensions.

We will not give an exhaustive review of this topic, but rather focus on some
minimal ingredients necessary for our following discussion. An interested reader
can find more details on the many geometric identities and on the relation with
String Theory in [24] and in references therein. There are three main types of
multiplets in $\mathcal{N} = 2$ supergravity: gravity, vector and hyper-multiplets

$$\begin{pmatrix} g_{\mu\nu} \\ \psi_\mu^A \\ A_\mu^0 \end{pmatrix}, \quad \begin{pmatrix} A_\mu^i \\ \lambda_A^i \\ z^i \end{pmatrix}, \quad \begin{pmatrix} \zeta^\alpha \\ q^u \end{pmatrix}, \tag{1.69}$$

$$\begin{array}{ccc} \text{gravity,} & n_V \text{ vector} & n_H \text{ hyper} \\ & \text{multiplets,} & \text{multiplets.} \end{array}$$

In a generic interacting ungauged theory, the number of vector fields is determined
by the number of vector multiplets n_V with the addition of the vector field sitting in
the gravity multiplet, named graviphoton. Scalar fields sit in both vector ($2n_V$ real
fields) and hypermultiplets ($4n_H$ real fields). The self-interactions between these
fields can be described by a factorized σ-model given by the product of a Special
Kähler manifold for the scalars in the vector multiplets and a Quaternionic Kähler
manifold for the scalars in the hypermultiplets:

$$\mathcal{M}_{\text{scalar}} = \mathcal{M}_{SK} \otimes \mathcal{M}_{QK}. \tag{1.70}$$

The different structure between the two manifolds has to do with the way the U(2)
R-symmetry group of $\mathcal{N} = 2$ theories acts on the fields of the two multiplets. Just
like the R-symmetry group factorizes U(2) = U(1) × SU(2), so does the scalar
manifold.

The scalars in the vector multiplets define the gauge kinetic functions $\mathcal{I}$ and $\mathcal{R}$,
while the hyperscalars do not enter into their definition. For this reason the scalars
of the hypermultiplets do not appear in the black hole potential V_{BH} and hence do
not participate at the definition of the tree level solutions. Hence, we will set them
to zero for the time being.

A Special-Kähler (SK) manifold can be parameterized by n_V complex scalar
fields, the scalars appearing in the vector multiplets. However, SK manifolds have
an intrinsic projective nature, related to the fact that there are $n_V + 1$ vectors
that appear in a supergravity theory that can mix between them. This means that
one could use $n_V + 1$ projective coordinates $X^{\Lambda(z)}$, $\Lambda = 0, 1, \ldots, n_V$, which are
holomorphic sections of the complex line bundle associated to the scalar manifold

(whose principal bundle is related to the U(1) factor in the R-symmetry group). Actually, just like for any set of $n_V + 1$ vector fields we can define a set of $n_V + 1$ duals according to the procedure outlined previously and symplectic duality transformations mix them, a SK manifold can be specified in terms of $2n_V + 2$ sections, which form a symplectic vector (X^Λ, F_Λ). Its Kähler potential is then defined via these sections as

$$K = -\log i \left(\bar{X}^\Lambda F_\Lambda - X^\Lambda \bar{F}_\Lambda \right) = -\log i \langle \Omega, \overline{\Omega} \rangle, \tag{1.71}$$

where

$$\langle A, B \rangle = A^T \begin{pmatrix} 0 & -1 \\ 1 & 0 \end{pmatrix} B. \tag{1.72}$$

Summarizing a *SK manifold is a Kähler manifold endowed with both a projective and a symplectic structure.*

Obviously the (X^Λ, F_Λ) sections and consequently the Kähler potential, is only defined locally. This means that, given two patches covering the scalar manifold U_α and U_β, the sections in their non-trivial intersection can be related by a symplectic and holomorphic transformation

$$\begin{pmatrix} X \\ F \end{pmatrix}_\alpha = S_{\alpha\beta}\, e^{h_{\alpha\beta}(z)} \begin{pmatrix} X \\ F \end{pmatrix}_\beta, \tag{1.73}$$

where $S_{\alpha\beta} \in \mathrm{Sp}(2n_V + 2, \mathbb{R})$ is constant. This implies a Kähler transformation on the Kähler potential

$$K_\alpha \to K_\beta + h_{\alpha\beta} + \bar{h}_{\alpha\beta}. \tag{1.74}$$

The projective nature becomes manifest in the fact that there is always a choice of the sections so that normal coordinates can be defined

$$t^i = \frac{X^i}{X^0}. \tag{1.75}$$

In such frames, the dual sections $F_\Lambda(z)$ can be derived from a prepotential $F(X)$, such that $F(\lambda X) = \lambda^2 F(X)$. We should stress that, on the other hand, generically there are frames in which such a prepotential does not exist.

The geometric structure of SK geometry fixes completely all the other couplings, among which the vector kinetic terms, which can be given in terms of a function $\mathcal{N}_{\Lambda\Sigma}$, with the property

$$F_\Lambda = \mathcal{N}_{\Lambda\Sigma} X^\Sigma. \tag{1.76}$$

The gauge kinetic couplings are the real and imaginary parts of this complex matrix:

$$R_{\Lambda\Sigma} = \mathrm{Re}\,\mathcal{N}_{\Lambda\Sigma}, \qquad I_{\Lambda\Sigma} = \mathrm{Im}\,\mathcal{N}_{\Lambda\Sigma}. \tag{1.77}$$

1.4.1 Examples

Before proceeding further, we give here a couple of interesting examples, which
will be used in the following.

The first example is one of the simplest SK manifold one could think of:
a manifold with a single scalar field parameterizing $SU(1,1)/U(1)$. In a frame where
a prepotential exists, it is defined as

$$F = -iX^0X^1, \tag{1.78}$$

which implies that $F_0 = -iX^1$ and $F_1 = -iX^0$. In such a frame we can also define
a normal coordinate $z = X^1/X^0$ and, with the gauge choice $X^0 = 1$, we can write
the holomorphic sections and the Kähler potential as

$$\Omega = \begin{pmatrix} 1 \\ z \\ -iz \\ -i \end{pmatrix}, \qquad K = -\log 2(z + \bar{z}). \tag{1.79}$$

The z modulus is constrained, because its real part must be positive in order for the
Kähler potential to be well defined.

A second simple example is the so-called STU model. This is a scalar manifold
corresponding to $[SU(1,1)/U(1)]^3$. The prepotential is

$$F = \frac{X^1X^2X^3}{X^0} \tag{1.80}$$

and the sections and Kähler potential can be written as

$$\Omega = \begin{pmatrix} 1 \\ s \\ t \\ u \\ -stu \\ tu \\ su \\ st \end{pmatrix}, \qquad K = -\log[-i(s - \bar{s})(t - \bar{t})(u - \bar{u})], \tag{1.81}$$

in a basis where $X^0 = 1$. It is interesting to point out that the metric of such manifold
factorizes.

1.4.2 String Theory Origin

Supergravity theories with $\mathcal{N} = 2$ supersymmetry in 4 dimensions can be obtained in various ways. The main path is to consider type II theories in 10 dimensions on Calabi–Yau threefolds. For instance, type IIB supergravity on a Calabi–Yau manifold Y_6 has $2(n_V + 1)$ three-cycles in $H_3(Y_6, \mathbb{R})$, with $n_V = h_{(2,1)}$, that lead to n_V vector multiplets in the effective theory by reduction of the Ramond–Ramond four-form of type IIB on them. The scalar fields in the corresponding SK manifold parameterize the space of complex structure deformations of the internal manifold. The holomorphic sections we introduced previously can be introduced by considering the periods of the holomorphic three-form Ω (with a suggestive abuse of notation) on the symplectic basis (A_Λ, B^Λ) of $H_3(Y_6, \mathbb{R})$:

$$X^\Lambda = \int_{A_\Lambda} \Omega, \qquad F_\Lambda = \int_{B^\Lambda} \Omega. \tag{1.82}$$

The corresponding Kähler potential can be obtained as

$$K = -\log i \int_{CY} \Omega \wedge \overline{\Omega}, \tag{1.83}$$

where the analogy between this expression, given in terms of the wedge product of the holomorphic three-form Ω, and (1.71), given in terms of the symplectic product of the sections Ω, is now evident.

By calling $(\alpha_\Lambda, \beta^\Lambda)$ the basis of harmonic three-forms on Y_6, the vector fields arise in the expansion of

$$F_5 = F^\Sigma \wedge \alpha_\Sigma - G_\Sigma \wedge \beta^\Sigma, \tag{1.84}$$

where the duality relation between G_Λ and F^Λ follows from the self-duality property of $F_5 = *F_5$. This expression is also telling us that the black hole charges in 4 dimensions correspond to charges of F_5 integrated over the product of a two-sphere and the three-cycles of Y_6. This means that black holes can be viewed as the superposition of D3-branes wrapping different three-cycles of Y_6, hence giving a hint on the route one needs to follow to explain the microscopic origin of the entropy of such configurations.

Since any Calabi–Yau manifold has at least a non-trivial three-cycle associated to the holomorphic form Ω, we can see why there is always at least one vector field in the corresponding $\mathcal{N} = 2$ effective theory, which appears in the gravity multiplet. The Kähler structure deformations are described by the hypermultiplet scalar fields.

Reductions of type IIA supergravity on a Calabi–Yau manifold are similar to the ones just described but with the role of complex and Kähler structure reversed: $\Omega \leftrightarrow J$. In particular, the vector-multiplet moduli space describes the complexified

Kähler structure of Y_6. If $J_c \equiv B + i\,J$, C_i is a basis of $H_{(1,1)}(Y_6, \mathbb{R})$ and D^i is the dual basis of $H_{(2,2)}(Y_6, \mathbb{R})$,

$$\frac{X^i}{X^0} = \int_{C_i} J_c, \qquad \frac{F_i}{F_0} = \int_{D^i} J_c \wedge J_c. \qquad (1.85)$$

The Kähler potential is now

$$K = -\log\left[\frac{4}{3}\int_{CY} J \wedge J \wedge J\right]. \qquad (1.86)$$

Vector fields generate from the 1, 3, 5 and 7-form potentials of type IIA expanded on the basis of harmonic 0, 2, 4 and 6-forms respectively. This means that the associated charges come from wrapped D0, D2, D4 and D6-branes.

1.5 Flow Equations for BPS and Non-BPS Attractors

In this section we are going to show that extremal black holes admit a first order description, no matter whether they are supersymmetric or not. For the sake of simplicity and in order to be specific, we will constrain our discussion to models within $\mathcal{N} = 2$ supergravity, but the results hold for more general theories. This presentation follows mainly [23], where the result was first derived, but expanding on the reasoning justifying and explaining it.

As we saw previously, critical points of the black hole potential define extremal black hole configurations and the same potential plays an essential role in the attractor mechanism. For $\mathcal{N} = 2$ theories the potential is

$$V_{BH} = |Z|^2 + 4g^{i\bar{j}}\partial_i|Z|\bar{\partial}_{\bar{j}}|Z|, \qquad (1.87)$$

where

$$Z = e^{K/2}\left(X^\Lambda q_\Lambda - p^\Lambda F_\Lambda\right) = e^{K/2}\langle\Omega, Q\rangle \qquad (1.88)$$

is the central charge of the $\mathcal{N} = 2$ supersymmetry algebra.

Extremal black holes are solutions of the equations of motion derived from the effective 1-dimensional lagrangian

$$\mathcal{L} = (U')^2 + g_{i\bar{j}}z^{i\prime}\bar{z}^{\bar{j}\prime} + e^{2U}\left(|Z|^2 + 4g^{i\bar{j}}\partial_i|Z|\bar{\partial}_{\bar{j}}|Z|\right), \qquad (1.89)$$

also satisfying the constraint

$$H = 0 \qquad \Leftrightarrow \qquad (U')^2 + g_{i\bar{j}}z^{i\prime}\bar{z}^{\bar{j}\prime} = e^{2U}\left(|Z|^2 + 4g^{i\bar{j}}\partial_i|Z|\bar{\partial}_{\bar{j}}|Z|\right) \qquad (1.90)$$

and where the scalar fields reach a critical point of the potential. The generic equations that need to be satisfied are second-order equations. We will now show that we can actually further reduce the system to first-order ordinary differential equations.

1.5.1 Supersymmetric Attractors

The Hamiltonian constraint (1.90) is an equality between two different sums of squares weighted with the same positive definite metric $g_{i\bar{j}}$. A natural solution is given by matching each term on the left hand side with the corresponding term on the right hand side as $U' = \pm e^U |Z|$ and $z^{i\,\prime} = \pm 2\, e^U g^{i\bar{j}}\bar{\partial}_{\bar{j}}|Z|$, for an arbitrary choice of sign in both equations. Although surprising at first sight, it is a straightforward exercise to show that such a solution of the constraint equation is also a solution of the equations of motion coming from (1.89), provided the same sign is chosen in the flow equations. We therefore reduced the system of second-order equations of motion and the quadratic constraint to a system of first order ordinary differential equations driven by the absolute value of the central charge $|Z|$.

The flow equation for the warp factor can also be rewritten as $(e^{-U})' = \mp |Z|$ and should be increasing along the flow, because its value is going to be 1 at infinity and becomes proportional to $|z|$ when approaching the horizon (see the discussion around (1.43)). This means that only the lower sign is acceptable in order to generate regular black hole solutions and hence black holes can be described by the following set of flow equations:

$$\begin{cases} U' = -e^U |Z|, \\ z^{i\,\prime} = -2\, e^U g^{i\bar{j}}\bar{\partial}_{\bar{j}}|Z|. \end{cases} \tag{1.91}$$

Having first-order equations rather than second-order, may be a sign of supersymmetry and in fact this is the case at hand. The mass of the black holes generated by (1.91) is

$$M_{ADM} = |Z|_\infty, \tag{1.92}$$

which means that they are extremal configurations at the threshold of the supersymmetric bound $M \geq |Z|$. In fact, by analyzing the gravitino and gaugino supersymmetry transformations one finds that, after imposing a suitable projector on the supersymmetry parameter, the first flow equation is equivalent to $\delta\psi^A_\mu = 0$, while the second satisfies $\delta\lambda^i_A = 0$. Actually, the scalar equation coming from the supersymmetry variation of the gauginos is

$$z^{i\,\prime} = -e^{U-i\alpha} g^{i\bar{j}}\overline{D}_{\bar{j}}\overline{Z}, \tag{1.93}$$

where α is a phase factor appearing in the projector, identified with the phase of the central charge. Full equivalence with (1.91) can be established by realizing that also the phase obeys a first order equation coming from the consistency of the supersymmetry conditions

$$\alpha' + Q = 0, \tag{1.94}$$

where Q is the composite Kähler connection $Q = \text{Im } z^{i\prime}\partial_i K$, and that this equation is identically satisfied once the flow equations (1.91) are fulfilled. This is an obvious consequence of the fact that α is not a new independent degree of freedom.

Inspection of (1.91) also shows that the central charge $|Z|$ determines completely the solution and that, no matter what is the value of the scalar fields at infinity, the flow stops where the central charged is minimized:

$$\partial_i |Z|_* = 0 \quad \Leftrightarrow \quad z^{i\prime} = 0. \tag{1.95}$$

As we could expect, such a critical point of the central charge is also a critical point of the full black hole potential V_{BH}:

$$\partial_i V_{BH} = |Z|\partial_i |Z| + \partial_i \partial_j |Z|\bar{\partial}^j |Z| + \partial_j |Z|\partial_i \bar{\partial}^j |Z| = 0 \tag{1.96}$$

and therefore a generic flow that reaches such a critical point describes a supersymmetric extremal black hole.

As expected, the attractor conditions (1.95) fix the values of the scalar fields in terms of the asymptotic charges of the solution $z^{i*} = z^{i*}(p,q)$ and all the horizon quantities depend only on the same charge values. The criticality condition (1.95) gives n_V complex independent conditions for n_V scalar fields and hence fixes them all.

At the critical point the warp factor has a simple behaviour

$$(e^{-U})' = |Z|_* \quad \Rightarrow \quad e^{-U} \to |Z|_* z, \tag{1.97}$$

so that the near horizon metric approaches $AdS_2 \times S^2$. Going back to the standard radial coordinate $r = -1/z$:

$$ds^2 = -\frac{r^2}{|Z|_*^2} dt^2 + \frac{|Z|_*^2}{r^2} \left[dr^2 + r^2 d\Omega_{S^2}^2\right]. \tag{1.98}$$

The corresponding black hole entropy is given by the usual area formula, which in this case can be rewritten in terms of the central charge and in turn of the black hole potential at the horizon:

$$S_{BH} = \frac{A}{4} = \pi |Z|_*^2 = \pi V_{BH}^*. \tag{1.99}$$

Since the scalar fields at the horizon are all fixed in terms of the electric and magnetic charges of the solution, also the central charge

$$|Z|_* = |Z|(p,q,z_*^i(p,q)) \tag{1.100}$$

depends only on the discrete charges and so does the entropy, according to (1.99).

Fig. 1.6 Representation of a
moduli space with multiple
basins of attraction

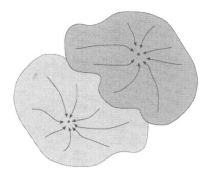

The geometric properties satisfied by the scalar manifold, namely the fact that
it has a Special-Kähler nature, constrain the evaluation of the second derivatives of
the central charge driving the supersymmetric flow so that

$$\partial_i \bar{\partial}_{\bar{j}} |Z| = g_{i\bar{j}} |Z| > 0. \tag{1.101}$$

This means that the critical points at which the flow stops are all minima of the
central charge. This also helps in understanding the attractor behaviour of the
black hole horizon. No matter what is the value at infinity of the scalar fields,
they are driven by the flow equations towards the minimum of the central charge,
which constitutes an attractor point for the differential equations determining the
flow. Eventually all the horizon quantities are determined by the value of the scalar
fields at such minimum. We can therefore think of our moduli space as a basin of
attraction where the final attractor point, at the minimum of the basin, is specified
only by the choice of the asymptotic charges.

We should note, however, that in some cases there can exist multiple basins of
attraction (Fig. 1.6), leading to a discrete number of possible values $z^i_*(p,q)$ for a
given choice of charges. In this case the attractor flow must be complemented by the
"area code" corresponding to the basin of attraction to which the initial conditions
belong
[62, 64, 81]. For a SK manifold all these critical points will be minima of the central
charge and there will be no other critical points, so that we are not in contradiction
with the previous discussion. If the reader wonders how such functions could be, an
example with two minima and no other critical point in $\mathbb{R}^2$ is $f = e^{-2x} - e^{-x-y^2} y^2$
[69].

Before proceeding to the analysis of the non-supersymmetric case, let us analyze
another interesting property of the flow equations (1.91): *the supersymmetric
c-theorem*. It is a straightforward consequence of the previous analysis that the scale
factor $\mu = e^{-U}$ is monotonically increasing along the flow

$$\mu' = |Z| > 0 \tag{1.102}$$

and therefore it can play the role of a c-function for the system (or rather its inverse), which has a minimum value at the Minkowski vacuum and blows up at the horizon (recall that $\mu_\infty = 1$ and $\mu_{hor} = +\infty$). This implies that μ can replace the radial coordinate to describe the same flow. By using the c-function as a parameter, the scalar field equations become a simple gradient flow equation:

$$\mu \frac{d}{d\mu} z^i = g^{i\bar{j}} \bar{\partial}_{\bar{j}} \log |Z|. \tag{1.103}$$

Here we focused on $N = 2$ models, but supersymmetric solutions in models with $N > 2$ follow essentially the same discussion where $|Z|$ is replaced by the largest of the absolute values of the eigenvalues of the central charge matrix.

1.5.2 Non-BPS Attractors

Although the solution of the Hamiltonian constraint given by (1.91) is straightforward and leads to supersymmetric attractors, the same theory (for different choice of charges) allows also for non-supersymmetric black holes. These are described by critical points of the scalar potential $\partial_i V^*_{BH} = 0$ for which the central charge is not minimized $\partial_i |Z|_* \neq 0$. The purpose of this section is to show that also in this case extremal black holes are described by first-order ordinary differential equations, driven by a function $W \neq |Z|$, which we will call fake superpotential [23].

There are two main motivations to believe that such a reduction may happen also for non-supersymmetric black holes. The first one is that the attractor mechanism is at work also in this case. Also for non-BPS extremal black holes the scalar potential drives the flow of the scalar fields in the moduli space towards the horizon value, which is once more specified by the minimum of the function. This is also generically an attractor point, in the sense that critical points of V_{BH} are generically minima, though for non-BPS black holes there exist flat directions. An interesting remark in this case is that such flat directions, when they exist, extend to the whole scalar potential and not just to the minimum. Also, derivative corrections do not seem to destabilize the model [10]. The second one is that also non-BPS attractors have a c-theorem and therefore the flow has a clear and unique direction of motion in the moduli space.

The proof of the non-BPS c-theorem was given in [57]. We leave the reader to the original reference for the precise demonstration, while here we recall the general line of the argument. The main assumption is that the matter involved in the theory giving the non-BPS extremal black holes satisfies the null energy condition. This states that the stress energy tensor for such models should be positive definite when contracted with null vector fields:

$$T_{\mu\nu} \zeta^\mu \zeta^\nu \geq 0, \qquad \forall \, \zeta \mid \zeta^2 = 0. \tag{1.104}$$

Once this assumption is made [57] proves that the area function A decreases monotonically along the solution towards the horizon and this means that one can identify such function with a c-function for the flow. The demonstration uses the fact that the area function is a specific term appearing in the metric, once again also proportional to e^{2U}, whose derivatives can be identified with certain combinations of the Ricci tensor constructed from the same metric: $A' \sim -R_{rr}g^{rr} + R_{tt}g^{tt}$. Now, using Einstein's equations and introducing a vector such that $(\zeta^t)^2 = -g^{tt}$ and $(\zeta^r)^2 = g^{rr}$, the same combination of the Ricci tensor describing the derivative of the area function can be rewritten as a combination of the stress energy tensor computed on the solution. Finally, using the null energy condition this shows that the area is always decreasing along the flow

$$A' \sim -R_{rr}g^{rr} + R_{tt}g^{tt} = -T_{\mu\nu}\zeta^{\mu}\zeta^{\nu} \le 0. \tag{1.105}$$

These facts and the analogy with a similar situation happening for non-BPS domain-wall solutions in the context of the gauge/gravity correspondence strongly suggests the existence of first-order equations also for non-BPS extremal black holes. Within the AdS/CFT correspondence Renormalization Group flows of the dual field theory can be described in the gravitational setup as domain-walls interpolating between two different Anti-de Sitter vacua. The field theory c-theorem guarantees a description of such domain-walls in terms of first order differential equations, also when there is no supersymmetry [22, 47, 75]. Analogously, extremal black holes are solutions interpolating between Minkowski and $AdS_2 \times S^2$ vacua of the same model. Using this analogy we can now make our argument solid.

Consider a simple model with a single scalar field ϕ, subject to a scalar potential $V(\phi)$ admitting two different extrema. An instantonic solution can be constructed by moving to euclidean signature and assuming that the field depends only on one variable, so that ϕ' denotes its derivative with respect to such variable. Different solutions are parameterized by different values of the energy and an extremal one is defined by solutions of the constrained equation of motion following from

$$\mathcal{L} = (\phi')^2 + V(\phi), \qquad H = (\phi')^2 - V(\phi) = 0. \tag{1.106}$$

Although this system has a generic second order equation of motion, it is easy to see that a first-order equation is sufficient by using Bogomolnyi's trick of squaring the action:

$$S = \int dt \left(\phi' \pm \sqrt{V} \right)^2 \mp 2 \int dt\, \phi' \sqrt{V}. \tag{1.107}$$

This action is equivalent to the one obtained by the previous Lagrangian, but now we clearly reduced the equation of motion for ϕ to a first order one

$$\phi' \pm \sqrt{V} = 0. \tag{1.108}$$

In fact, the second term in (1.107) is always a total derivative and hence can be discarded, while the first term in brackets solves both the equation of motion and the Hamiltonian constraint.

The extension of this trick to the case where many scalars ϕ are involved needs some care. Extremal solutions of a system analogous to (1.106) but with many scalars means solving the constrained equations of motion coming from

$$\mathcal{L} = |\phi'|^2 + V(\phi), \qquad H = |\phi'|^2 - V(\phi) = 0, \tag{1.109}$$

where the norm $|\phi'|$ can be taken with respect to a positive definite metric also depending on the scalar fields. The squaring of the action leads to

$$S = \int dt \left| \phi' \pm n \sqrt{V} \right|^2 \mp 2 \int dt \, n \cdot \phi' \sqrt{V}, \tag{1.110}$$

where n is a unit-norm vector: $|n|^2 = 1$. In this example however the second term in the action is a boundary if and only if it is proportional to the field derivative of a new function W:

$$n = \frac{\nabla_\phi W}{\sqrt{V}}. \tag{1.111}$$

We therefore conclude that the system of equations of motion and Hamiltonian constraint coming from (1.109) can be described by first-order equations provided the scalar potential can be rewritten as the norm of the derivative of a scalar function:

$$V(\phi) = |\nabla_\phi W|^2. \tag{1.112}$$

The Lagrangian and the Hamiltonian constraint of extremal black holes are a special instance of this general case, where the set of scalars comprises the moduli fields as well as the warp factor $\phi = \{U, z^i\}$ and the metric defining the norm is factorized and equal to 1 in the U direction and equal to $g_{i\bar{j}}$ in the direction of the moduli fields. Actually, given the special dependence on the warp factor, we can introduce a real valued *fake superpotential* W so that $\mathcal{W} = e^U W$ and the necessary constraint to rewrite the equations of motion in a first-order form reduces to

$$e^{2U} V_{BH} = \partial_U (e^U W)^2 + 4 \, \partial_i (e^U W) g^{i\bar{j}} \partial_{\bar{j}} (e^U W), \tag{1.113}$$

which implies that the black hole potential can be written as [23]:

$$V_{BH} = W^2 + 4 \, \partial_i W g^{i\bar{j}} \partial_{\bar{j}} W. \tag{1.114}$$

Whenever this condition is satisfied, the black hole equations are reduced to first-order differential conditions [23]:

$$\begin{cases} U' = -e^U W, \\[2mm] z^{i\,\prime} = -2 \, e^U g^{i\bar{j}} \, \partial_{\bar{j}} W. \end{cases} \tag{1.115}$$

Once more the sign is fixed because of consistency of the behaviour of the warp factor from infinity to the horizon. Obviously the BPS case is trivially recovered whenever $W = |Z|$, however, we will see that for a given black hole potential more solutions, with $W \neq |Z|$, can and will exist whenever there are extremal points of the black hole potential that are not extrema of the central charge.

The fake superpotential W assumes a very important role for the description of the black hole solutions. In fact, not only the flow equations determining the complete solution are driven by W, but also the mass and the entropy of the black hole are determined by the same function [43]. In detail, the boundary term reduces to $e^U W$ and the value of the boundary term far away from the black hole determines the ADM mass: $e^U W \to W_\infty = M_{ADM}$. Also in this case the flow equations will stop at the critical point of the function driving the scalars. For non-BPS black holes this means that the horizon is reached whenever

$$\partial_i W_* = 0. \tag{1.116}$$

It is trivial to check that these are also critical points of the full black hole potential V_{BH}. As discussed previously, in the non-BPS case there may be flat directions and actually this is reflected by W, which will not depend on the moduli related to such flat directions [3, 25, 27]. From the flow equations we can also determine the behaviour of the warp factor close to the horizon, in full analogy with the supersymmetric case (1.97), and this fixes the entropy to be

$$S_{BH} = \frac{A}{4} = \pi W_*^2 = \pi V_{BH}^*. \tag{1.117}$$

We stress that the fake superpotential is an extremely powerful procedure that provides the full solution, including properties that depend on the behaviour of the scalar fields infinitely far away from the black hole, and not just the horizon properties as other procedures do.

Obviously the main problem connected with this technique is whether we can find any solution of the main constraint equation (1.114) other than the central charge of the system. The answer is positive and two main techniques have been developed to provide such answer:

• A constructive approach for coset manifolds based on duality invariants (see [25, 27] and [2] also for $N > 2$ theories);
• An existence theorem in connection with the Hamilton–Jacobi equation [4].

The constructive approach is based on the simple observation that the warp factor U is a duality invariant quantity (it is part of the metric and this is invariant under U-duality transformations). Since the derivative of this function is related to W, also the fake superpotential must be invariant. For a model based on a symmetric coset manifold describing the self-interactions of the vector multiplet scalar fields $\mathcal{M}_{sc} = G/H$, the duality group is $G \subset \mathrm{Sp}(2n_V + 2, \mathbb{R})$. It is therefore a straightforward technical task to identify all possible invariants and find a working definition for W. This can generically be done in few steps, by first identifying W for a simple charge configuration, using symmetry properties to reconstruct the seed superpotential and then boost it by a duality transformation to generic charges. One has to say that although the number of invariants is limited and one generically faces a well defined problem, the resulting fake superpotential may be extremely non-trivial, for instance non-polynomial in the invariants (see [11]) and hence the procedure can be technically challenging. Also, this procedure does not apply to the cases where the scalar manifold is not a coset.

On the other hand, the Hamilton–Jacobi interpretation of our system of equations is very useful to derive a formal general solution and to prove an existence theorem, but it is also often unpractical in order to derive a closed expression for W. The Hamilton–Jacobi equation is a first order nonlinear partial differential equation for a function $\mathcal{W}(\phi)$ called Hamilton's principal function such that

$$\mathcal{H}\left(\phi^i, \frac{\partial \mathcal{W}}{\partial \phi^i}\right) = 0 \qquad (1.118)$$

In our case the radial variable plays the role of a euclidean time, the fake superpotential plays the role of the principal Jacobi function while the set of all fields ϕ^i is assimilated to the coordinates of phase-space and the equation to be solved is defined by

$$\mathcal{H} = V(\phi, Q) - \left|\frac{\partial \mathcal{W}}{\partial \phi}\right|^2. \qquad (1.119)$$

In this context $\mathcal{W}$ gets the interpretation of the generating function of canonical transformations of the classical Hamiltonian, so that $\pi^i = \frac{\partial \mathcal{W}}{\partial \phi^i}$ and $\pi^i = G_{ij}\phi^{j\prime} = \frac{\delta \mathcal{L}}{\delta \phi^{i\prime}}$ become the flow equations. The existence of such a function is guaranteed by the Liouville integrability of the system [46]. From the general theory, $\mathcal{W}$ can also be formally constructed as the integral of the Lagrangian along the solution [4]:

$$\mathcal{W}(\phi) = \mathcal{W}_0 + \int_{\tau_0}^{\tau} \mathcal{L}(\phi, \phi')d\tau. \qquad (1.120)$$

Clearly this is unpractical in the generic situation where one does not know the solutions before having constructed $\mathcal{W}$.

An alternative technique has been proposed in [19], where the superpotential is implicitly defined via the solution of an algebraic equation of degree 6, which follows from analyzing the geodesics in the time-like reduction of the black hole geometry. This also gives a formal general definition of W, but the result of the solution of the equation of degree 6 is at least impractical.

1.5.2.1   Examples

The first example is given by the SU(1,1)/U(1) model with prepotential

$$F = -iX^0 X^1. \tag{1.121}$$

The holomorphic sections and the Kähler potential were given in Sect. 1.4.1. For generic electric q_Λ and magnetic p^Λ charges, the central charge is then

$$Z = \frac{q_0 + ip^1 + (q_1 + ip^0)z}{\sqrt{2(z + \bar{z})}}. \tag{1.122}$$

The black hole potential V_{BH} is derived by inserting this expression in (1.87):

$$V_{BH} = \frac{(p^1)^2 - iq_1(z - \bar{z})p^1 + q_0{}^2 + ip^0 q_0(z - \bar{z}) + ((p^0)^2 + (q_1)^2)z\bar{z}}{z + \bar{z}}. \tag{1.123}$$

Black hole solutions are then found by looking for solutions interpolating between flat space at infinity and $AdS_2 \times S^2$ at the horizons defined by the critical points of V_{BH}. Such critical points are found for

$$z^\pm = \frac{\pm(p^0 p^1 + q_0 q_1) + i(p^0 q_0 - p^1 q_1)}{(p^0)^2 + (q_1)^2}, \tag{1.124}$$

and since consistency requires Re$z > 0$, they lie inside the moduli space only for $(p^0 p^1 + q_0 q_1) > 0$ when z^+ is chosen in (1.124), and for $(p^0 p^1 + q_0 q_1) < 0$ for z^-. Different critical points have a different nature. More precisely, z^+ (1.124) gives the supersymmetric vacuum, which satisfies $D_i Z = 0$, with $Z \neq 0$, (hence $\partial_i |Z| = 0$) and thus it is a fixed point of (1.122), while the other critical point z^- gives the non-BPS black hole, for which $D_i Z \neq 0$. The Hessian at these points is always positive as there are two identical positive eigenvalues

$$\text{Eigen}\{\text{Hess}(V_{BH})\} = \pm \frac{1}{p^0 p^1 + q_0 q_1}\{((p^0)^2 + (q_1)^2)^2, ((p^0)^2 + (q_1)^2)^2\}. \tag{1.125}$$

A simple inspection of these formulae shows that the two type of black holes are related by a change of sign in the electric or magnetic charges. In fact, in this case, the fake superpotential is given by

$$\mathcal{W} = \frac{|-q_0 + ip^1 + (q_1 - ip^0)z|}{\sqrt{2(z + \bar{z})}}, \tag{1.126}$$

which indeed differs from (1.122), but gives rise to the same potential V_{BH}. It is also quite simple to check that the critical point of this new "fake superpotential" is the non-BPS black hole, namely (1.124) with the minus sign (Fig. 1.7).

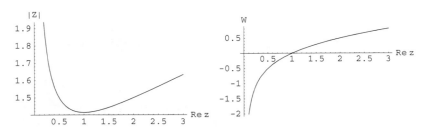

Fig. 1.7 Plots of the sections of W and $|\mathcal{Z}|$ at Im $z = 0$, for unit charges. Where the central charge shows a minimum, the "fake superpotential" crosses zero. Changing the signs of the q_0 and p^0 charges exchanges the two pictures

An extremely non-trivial example is the STU model, for which the complete superpotential expression can be found in [11]. A simpler expression we can report here is the limiting case of the t^3 model, which essentially follows from the STU model by identifying $s = t = u$, for the case of only two non-trivial charges, p^0 and q_1. The central charge is

$$Z = \frac{zq_1 + p^0 z^3}{\sqrt{-i(z - \bar{z})^3}}, \tag{1.127}$$

whose critical point is

$$z^* = -i\sqrt{-\frac{q_1}{3p^0}}. \tag{1.128}$$

The corresponding fake superpotential is

$$W = \frac{|zq_1 + p^0 z^2 \bar{z}|}{\sqrt{-i(z - \bar{z})^3}}, \tag{1.129}$$

whose critical point is

$$z^* = -i\sqrt{\frac{q_1}{3p^0}}. \tag{1.130}$$

This superpotential cannot be obtained from the central charge just by flipping charges. Note also that when the supersymmetric critical point is well defined, the non-BPS critical point is not well defined and vice-versa.

1.6 Duality

An important aspect of black hole solutions in supergravity is that the U-duality transformations map solutions of the various equations of motion and Bianchi identities to new solutions of the equations of motion and Bianchi identities of another system with different charges, preserving the metric. This implies that the U-duality group U can be used to generate solutions with arbitrary charges starting from more constrained configurations (see for instance $[1, 28–30, 42, 60, 61, 71, 74]$) by

$$\begin{pmatrix} p' \\ q' \end{pmatrix} = S \begin{pmatrix} p \\ q \end{pmatrix},$$ (1.131)

where $S \in U \subset Sp(2n_V + 2, \mathbb{R})$. We have also seen that such a technique has been used in order to provide a constructive mechanism to find the fake superpotential in the case of coset scalar manifolds.

An obvious interesting question is: what is the minimum number of charges that allows to generate arbitrary ones by a duality transformation? The black hole solutions we considered so far are specified by the values of the charges (p^Λ, q_Λ) and by the asymptotic value of the scalar fields $z^i_{\omega=+\infty}$. This gives a total of $2n_V + 2 + 2n_V$ parameters specifying the complete solutions. Note that while the first set is enough for defining the various quantities at the horizon, the second set is necessary for the full solution. Consider now the case where the scalar manifold is given by a coset $\mathcal{M}_{sc} = G/H$. The asymptotic values of the scalar fields $z^i_{\omega=+\infty}$ are parameters of the coset manifold and hence we can use the generators of G/H to set them to whatever value we want. The non-trivial parameters left are then $2n_V + 2$, corresponding to the electric and magnetic charges of the theory. We are also free to use duality transformations sitting in H, because they will not affect the value of the scalar fields,[2] but will rotate the charges. Some H-generators may also have a trivial action on the charges, but all the others can be used to remove parameters of the solutions, which could be generated by the action of the duality group, instead.

An explicit example of this procedure can be outlined for the STU model. A generic black hole in this model has 14 parameters: 8 charges and 6 real scalar fields. The duality group is determined by the scalar manifold to be $G = SU(1,1)^3$, with $H = U(1)^3$. As expected, the dimension of the coset coincides with the number of scalar fields and therefore the minimum number of parameters can be obtained by subtracting the dimension of H to the total number of charges

$$\#charges - \#(H) = 8 - 3 = 5.$$ (1.132)

A slightly more complicated example is given by $N = 8$ supergravity. This theory has 56 charges and 70 scalar fields for a total of 126 parameters for the generic black hole configuration. The scalar manifold is $E_{7(7)}/SU(8)$, and the dimensions of the two groups are $\#[E_{7(7)}] = 133$ and $\#[SU(8)] = 63$. Surprisingly, if we straightforwardly apply the previous relation and subtract the number of H-generators to the number of allowed charges we would get a negative number.

---

[2]To be more precise: the scalar manifold can be parameterized by (right-invariant) coset representatives L, from which one can construct the scalar matrix $M = L^T L$ that enters in the scalar kinetic term. M is invariant under H-transformations.

However, the resolution of this little puzzle is rather simple if one analyzes more carefully the action of the H-group on the same charges. In fact there is an $SU(2)^4 \subset H$ that leaves the charges invariant and therefore the total number of duality parameters we can use to reduce the number of independent charges is $\#(H) - \#[SU(2)^4] = 51$. This means that we are once more left with just five independent parameters to construct the seed solution that can be used to generate the most general one by duality. This is not surprising after all, because the STU model analyzed previously is a special truncation of the $N = 8$ model. It should also be noted that, although we fixed the scalars in order to determine the number of independent parameters, the seed solution could also be constructed by choosing any other five parameters among the charges and asymptotic scalars, with the obvious constraint that the invariant charge combination defining the entropy at the horizon should be non-vanishing (for large black holes).

Another important issue we will not analyze here in detail is the definition of the duality orbits [1, 8, 42]. Since duality transformations can map black holes with different charges among themselves, it is useful to understand how many different orbits one has with respect to this action. A full classification requires the construction of an appropriate number of independent invariant quantities. For instance in $N = 8$ supergravity different orbits are classified by the quartic invariant $I_4(p, q)$. Whenever $I_4 > 0$ one has supersymmetric black holes. For $I_4 < 0$ one has non-BPS configurations and, finally, when $I_4 = 0$, the horizon area vanishes and hence one has small black holes. $N = 2$ duality orbits were classified in [26].

1.7 Multicentre Solutions

So far we concentrated our discussion on single centre black hole configurations. However, a great deal of progress in our recent understanding of black hole physics within the context of String Theory came from multicentre solutions. In this final section we will review such solutions with a special emphasis on the non-supersymmetric ones.

Once more, we would like to make contact with what has been discussed so far in the context of single centre configurations and therefore we will focus on $N = 2$ theories in 4 dimensions. In order to have an explicit relation with String Theory it is actually very useful to see also how such configurations can be constructed in 10/11 dimensions within type II or M-theory models. We chose to concentrate on M-theory compactifications on the product of a Calabi–Yau manifold and a circle: $CY \times S^1$. This kind of compactification leads to an $N = 2$ theory in 4 dimensions and it is also useful for to establish an explicit relation between quantities in 4 and 5 dimensions. We will further specify the Calabi–Yau to be a simple orbifold

$$CY_6 = \frac{T^6}{\mathbb{Z}_2 \times \mathbb{Z}_2} \simeq (T^2)^3, \qquad (1.133)$$

because this gives a reduced setup where the scalar manifold is described by the STU model. In this scenario the eight charges of the STU model correspond to different branes wrapped on the various cycles of the internal manifold. Obviously, by replacing the internal manifold with a more general Calabi–Yau, we can get more general cubic Special–Kähler geometries, as discussed in Sect. 1.4.2. If we call t and x the coordinates of 4-dimensional spacetime, ψ the coordinate of the circle and y^a the coordinates of the six-torus, with the two $\mathbb{Z}_2$ orbifold actions inverting the sign of the first four and last four coordinates respectively, the charge configuration can be summarized by the following table:

Dimensions wrapped by the various M-branes and D-branes with the corresponding 4-dimensional charges

M-theory	t	x_1	x_2	x_3	ψ	y_1	y_2	y_3	y_4	y_5	y_6	IIA	Charge
$KK6$	—				—	—	—	—	—	—	—	$D6$	p^0
$M5$	—				—	—	—	—	—			$D4$	p^1
$M5'$	—				—	—	—			—	—	$D4'$	p^2
$M5''$	—				—			—	—	—	—	$D4''$	p^3
$M2$	—					—	—					$D2$	q_1
$M2'$	—							—	—			$D2'$	q_2
$M2''$	—									—	—	$D2''$	q_3
$KK0$	—				—							$D0$	q_0

where the line denotes the extension of the brane along that direction. From the M-theory point of view we have six real charges (p^i and q_i) and two geometric charges (p^0 and q_0). The first one arises from Kaluza–Klein monopoles in M-theory, and can be seen as an additional brane charge in 10 dimensions (a D6-brane charge), while q_0 arises as the charge related to the 4-dimensional vector arising from the 5-dimensional component of the metric describing the fibration of ψ on x and corresponds to a Kaluza–Klein particle (with a nontrivial momentum along ψ).

It is actually better to describe the reduction process to 4 dimensions in two steps. First of all we consider the reduction to 5 dimensions along the $(T^2)^3$ and then discuss further reductions to 4 dimensions. From the M-theory point of view a stationary metric ansatz that takes into account the backreaction of the above configuration of branes on the geometry is

$$ds_{11}^2 = -Z^{-2}\,(dt+\omega)^2 + Z\,ds_4^2(x) + \sum_I \frac{Z}{Z_I}\,ds_{T^2}^2. \qquad (1.134)$$

Here ds_4^2 is a Ricci-flat 4-dimensional euclidean space, which can be chosen to be $\mathbb{R}^4$, the Gibbons–Hawking space, the Euclidean Schwarzschild metric or others according to the type of solution we want to describe. The rest of the ansatz is chosen so that the total volume of the internal manifold remains fixed, which implies that no hypermultiplets will be turned on in the lower-dimensional effective theory. The warp factors also depend only on the coordinate of the 4-dimensional euclidean space and

$$Z = (Z_1 Z_2 Z_3)^{1/3}. \tag{1.135}$$

Finally, the M-theory ansatz is completed by the three-form potential, which we take as

$$C_3 = \sum_I \left(\frac{-dt + \omega}{Z_I} + a_I \right) \wedge dT_I, \tag{1.136}$$

where dT_I is the volume element of the I-th torus. This gives the so-called *floating brane ansatz*. The name follows from the fact that any probe M2 brane wrapping one of the two-tori feels no force thanks to the cancellations between the contributions coming from the Dirac–Born–Infeld action and the Wess–Zumino terms due to the same dependence of the metric and C_3 on the warp factors Z_I.

By using this ansatz, the Bianchi identities and equations of motion of the 11-dimensional theory reduce to almost linear equations depending only on the coordinates of $ds_4^2(x)$ [12]:

$$da_I = \star_4 da_I, \tag{1.137}$$

$$d \star_4 dZ_I = \frac{|\epsilon_{IJK}|}{2} da_J \wedge \star_4 da_k, \tag{1.138}$$

$$d\omega + \star_4 d\omega = Z_I da_I. \tag{1.139}$$

As discussed above, in the end we would like to describe black hole geometries in 4 dimensions. This means that we need to choose the metric of ds_4^2 as a Ricci-flat circle fibration on a 3-dimensional base, so that we can reduce the model along the circle direction and go back to the STU model. The first choice we will consider is that of a Gibbons–Hawking space. This is a 4-dimensional euclidean space endowed with a metric

$$ds_4^2 = \frac{1}{V} (d\psi + A)^2 + V dx_3^2, \tag{1.140}$$

where

$$\star dA = \pm dV. \tag{1.141}$$

We assume that ψ is a U(1) isometry and hence V depends only on the coordinates of the 3-dimensional flat base $dx_3^2 = dx_1^2 + dx_2^2 + dx_3^2$. By consistency, V is a harmonic function which generates the NUT charge corresponding to the presence of a D6 brane in 10 dimensions:

$$V = h^0 + \frac{p^0}{r}. \tag{1.142}$$

The overall geometry of this space is that of a cigar (Look at Fig. 1.8). At the tip $(r \to 0)$ one has the NUT charge p^0 and spacetime looks 5-dimensional. At large

Fig. 1.8 BH's in a GH
geometry

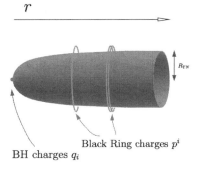

Black Ring charges p^i

BH charges q_i

values of r, the Gibbons–Hawking space becomes the direct product of $\mathbb{R}^3$ with a
circle of radius

$$R_{TN} = \frac{1}{\sqrt{h^0}}. \tag{1.143}$$

The choice of sign in (1.141) is equivalent to a choice of orientation on the space
and has a dramatic effect on the reduction of the equations of motion (1.137)–
(1.139). Depending on the sign one gets two different sets of equations which
correspond to supersymmetric and non-supersymmetric configurations [14, 56].
In detail, the plus sign gives BPS configurations, whereas the minus sign breaks
supersymmetry. The supersymmetry breaking is obviously mild and is essentially
just a reaction to the change in the global conditions (the orientation of the space).
For this reason these conditions have been named *almost BPS* [56].

The knowledge of the map between the quantities entering in the definition of the
11-dimensional ansatz and the fields defining the 4-dimensional STU supergravity
model allows the use of the 4-dimensional duality group to generate new solutions
starting from known ones. The full map is given in [32]. However, we show here how
the reduction along the ψ circle allows the definition of new quantities in terms of
which one defines the physical scalars in 4 dimensions. Now quantities with a vector
have legs only along $d\mathbf{x}_3^2$. The one-forms appearing in the metric and M-theory
potential now split as

$$a_I = C^I (d\psi - A^0) + A^I, \tag{1.144}$$

$$\omega = \mu(d\psi - A^0) + \boldsymbol{\omega}, \tag{1.145}$$

so that the scalar fields of the STU model are parameterized by

$$z^I = C^I - i\frac{X^I}{\Delta^2}, \tag{1.146}$$

where $X^I = Z/Z_I$ is the volumes of the I-th torus and Δ is a factor that depends on the fibration. For the Gibbons–Hawking reduction [32]

$$\Delta^4 = \frac{Z^{2/3}V}{VZ^3 - V^2\mu^2}. \tag{1.147}$$

As explained in Sect. 1.4.1, each coordinate spans a SU(1,1)/U(1) factor and the U-duality group is $[SU(1,1)]^3$. The action of the duality group on the scalar fields can be represented as a fractional linear transformation

$$z^i \rightarrow \frac{a_i z^i + b_i}{c_i z^i + d_i}, \tag{1.148}$$

where the parameters are part of an SU(1,1) valued matrix:

$$M_i = \begin{pmatrix} a_i & b_i \\ c_i & d_i \end{pmatrix} \in SU(1,1)_i. \tag{1.149}$$

We stress that each of these transformations acts also on the charges, but leaves the 4-dimensional metric invariant. On the other hand, such transformations do not leave the 5-dimensional or 11-dimensional metric invariant and this may have profound consequences on the form of the solution as seen from M-theory. Although some of the 4-dimensional duality transformations have an obvious 11-dimensional interpretations, other become rather non-trivial and involved in the uplifting process. For instance, the combination of 2 T-duality transformations on the I-th torus gives a matrix transformation of the form

$$M_I = \begin{pmatrix} 0 & -1 \\ 1 & 0 \end{pmatrix} \tag{1.150}$$

on the corresponding scalar. It is interesting to note that from the 4-dimensional point of view this is an S-duality transformation $z \rightarrow -1/z$. Gauge shift symmetries of C_3 along the tori also have a straightforward representation:

$$M_I = \begin{pmatrix} 1 & \lambda_I \\ 0 & 1 \end{pmatrix}. \tag{1.151}$$

On the other hand, the action of

$$M_I = \begin{pmatrix} 1 & 0 \\ \lambda_i & 1 \end{pmatrix} \tag{1.152}$$

has the interpretation of a spectral flow transformation and only the rewriting in terms of 4-dimensional duality transformations allowed us to identify such a transformation with the combination of 6 T-dualities, a gauge transformation and 6 inverse T-dualities. The same duality transformations act also on the charge vectors as described in [6, 32].

1.7.1 BPS Case

The 4-dimensional black hole solution is generically described by the warp factor U, by the scalar fields z^i and by the time components of the electric and magnetic vector fields $\{A^\Lambda, A_\Lambda\}$, whose duals we replaced with the charges in previous examples. In the multi-centre case, the solutions will not be static anymore and therefore one also needs the one-form ω. Using the setup considered in this section, the electric vector fields follow from the reduction as $A^\Lambda \sim \{A^0, A^I\}$.

In the case of the choice of positive orientation on the Gibbons–Hawking metric, the solution is supersymmetric. The full solution was first found in [34] and it can be completely expressed in terms of eight harmonic functions

$$\star_3 d\, A^\Lambda = dH^\Lambda, \tag{1.153}$$

$$\star_3 d\, A_\Lambda = dH_\Lambda, \tag{1.154}$$

with

$$H = h + \sum_i \frac{Q_i}{|x - x_i|}, \tag{1.155}$$

where Q_i represent the appropriate electric $q_{\Lambda i}$ or magnetic charge p_i^Λ. Also the non-static part of the metric is given in terms of the same harmonic functions

$$\star\, d\omega = \langle dH, H \rangle \tag{1.156}$$

and its existence is related to the intrinsic angular momentum due to the electric-magnetic field generated by the static charges.

Consistency also implies that the positions of the charges is constrained by the equations

$$\sum_j \frac{\langle Q_i, Q_j \rangle}{|x_i - x_j|} = 2\mathrm{Im}\,(e^{-i\alpha} Z(q_i))_\infty, \tag{1.157}$$

whose zeros give the positions in moduli space related to marginal stability of the solution (some of the distances blow up and therefore components are not bound anymore).

The application of duality transformations to this solution simply rotates the harmonic functions among themselves [32]. This means that for a supersymmetric solution the 11-dimensional metric and three-form potential are always described by an ansatz like the one we presented above, though the details in terms of the charges depend on the choices of the harmonic functions.

1.7.2 Non-BPS Case

The non-supersymmetric case is more interesting (at least for what concerns the construction of solutions). In this case not all possible solutions will fall in the ansatz above. Actually many interesting new solutions that are candidate microstate geometries for black holes can be obtained by different bubbling equations. These depend on the form of our choice of ds_4^2 and consequently on the choice of the three-form potential and warp factors. Starting from the Gibbons–Hawking space (Almost BPSsolutions), the bubbling equations become

$$dA^I = C^I dV - VdC^I, \qquad (1.158)$$

$$d \star_3 dZ_I = \frac{|\epsilon_{IJK}|}{2} Vd \star_3 d(C^J C^K), \qquad (1.159)$$

$$\star_3 d\omega = d(\mu V) - VZ_I dC^I. \qquad (1.160)$$

In opposition to what happened in the case of supersymmetric solutions one cannot generally solve these equations only in terms of harmonic functions. If one insists in doing so, then only mutually local solutions exist and the positions of the various centers are not constrained [49, 55]. However one can find more general solutions with mutually non-local charges and constrained positions.[3]

Before moving to the multi-centre case, we can see that from the above ansatz one can easily recover the single centre seed solution with D2 and D6 charges and a non-trivial axion [14, 54, 66] (this is a total of four charges and one non-trivial asymptotic value for the scalar fields). This corresponds to setting $C^I = 0$. The bubbling equations imply that now V and Z_I are harmonic functions and $\mu = \frac{b}{V}$, where b is the asymptotic value of the axions. The warp factor of the black hole metric follows as

$$e^{-4U} = 4VZ_1Z_2Z_3 - b^2. \qquad (1.161)$$

---

[3]The apparent discrepancy between the results in [49, 55] on the one hand and [14] on the other is due to the different implementation of the regularity requirements. In order to have regular horizons, the warp factor should not grow too fast when approaching the horizon. If this behaviour is constrained by considering the asymptotic charges, the only possible regular solutions are all marginally stable and the solution is given entirely in terms of harmonic functions. On the other hand this global requirement is not necessary and application of this condition centre by centre allows for regular bound states as in [14].

For a multicentre solution we can start by requiring C^I to be harmonic functions. This, in turn, implies that the Z_I are not harmonic anymore. The full solution is rather complicated and can be found in [14]. We stress that regularity of the solution in this case implies the existence of non-trivial constraints already for two centres. This is actually an equation for the distance between the centres that is given in terms of charges and asymptotic values of the scalar fields.

The existence of a constraint on the distances implies that such solutions describe bound states of black holes. It is obviously of much interest to understand the general conditions under which such bound states exist depending on the charges and the point in moduli space, like in the BPS case, where such an analysis culminated in the wall crossing formulae (See for instance the lectures at [35, 70]). A first attempt at an analysis in this direction is presented in [44, 45], but only in special models and for two-centre solutions.

Once more the solution obtained in [14] can be used as a seed to obtain more general ones by using duality transformations. However, in this instance, the action of the duality group changes the 5-dimensional metric [32]. We show here a simple example. Choose $C^1 = C^2 = 0$ and $C^3 \neq 0$, which means only one non-trivial D4-charge. By inspection of the bubbling equations we now see that the 11-dimensional warp factors Z_I are harmonic and only μ cannot be given in a closed form in terms of harmonic functions, because of its equation, which is

$$d \star d(V\mu) = d(VZ_3) \wedge \star dC^3. \tag{1.162}$$

Performing 6 T-dualities, following the approach described in the previous section, the new 5-dimensional metric contains a new spatial part, which describes now an Israel–Wilson space, rather than a Gibbons–Hawking. The dual metric is

$$ds_4^2 = (V_1 V_2)^{-1}(d\psi - A)^2 + V_1 V_2 \, dx_3^2, \tag{1.163}$$

where $V_1 = C^3$, $V_2 = Z_3$ and $\star dA = V_2 \, dV_1 - V_1 \, dV_2$. It is obvious that this will never be of the Gibbons–Hawking form.

The fact that the floating brane ansatz encompasses all BPS solutions, whereas it does not contain all non-BPS solutions, implies that the non-supersymmetric configurations are far more richer than the supersymmetric ones. Unfortunately, this also means that the attempts at constructing a general solution, valid in any frame, as was done for the BPS multicentre black holes in [34], have to face harder challenges. Some progress in a 4-dimensional setup was made in [50]. Moreover, having more centres means that duality orbits are classified by a larger number of invariants [45] and the non-BPS ones of [14] are the seed solutions only for one of the orbits in [45].

Similar techniques can be used to generate more general new classes of non-BPS solutions, however it is very difficult to find a sufficiently general explicit solution that fulfills the constraint on the positions. The best obtained so far was a line of rotating black holes [15].

Before wrapping up, let us also mention that such bubbling equations are also very useful to obtain other general solutions whenever ds_4^2 has a more general form. In particular one can obtain new 5-dimensional smooth solutions with the same charges of 4-dimensional black holes and hence interpret these states as candidate microstate geometries [16–18].

Also, the reduction to 4-dimensions of these stationary solutions allows also for rotating configurations in 4-dimensions as well. In fact, by this mechanism we could find the most general seed solution for slowly rotating black holes ($a \to 0, m \to 0$, $J = a/m$ fixed) [14]. The overrotating counterpart ($a = m \neq 0$) is still missing because of the different structure of the ansatz.

As a final note we would like to note that the idea of reducing extremal black hole equations to a first-order formalism [23] has revised the search for exact solutions also in other instances where some of the conditions assumed in [23] are relaxed. Recently we have seen this formalism applied to supersymmetric black holes in U(1) gauged supergravity [31] (solutions in this context were also obtained in [9,21,59]), to rotating extremal solutions [50] and even to classes of non-extremal black holes [4,51,72].

Acknowledgements The original parts of the contents of these lectures come from collaborations with I. Bena, A. Ceresole, S. Ferrara, S. Giusto, G. Lopes Cardoso, J. Perz, C. Ruef, C. Toldo, A. Yeranyan and N. Warner, which are gratefully acknowledged. I would also like to thank the organizers of the "27th Nordic Spring String Meeting", "SAM 2009" and "BOSS 2011" schools for the kind hospitality and the nice and stimulating environment. This work is supported in part by the ERC Advanced Grant no. 226455, *"Supersymmetry, Quantum Gravity and Gauge Fields" (SUPERFIELDS)*, by the Fondazione Cariparo Excellence Grant *String-derived supergravities with branes and fluxes and their phenomenological implications*, by the European Programme UNILHC (contract PITN-GA-2009-237920) and by the Padova University Project CPDA105015/10.

References

1. L. Andrianopoli, R. D'Auria, S. Ferrara, U duality and central charges in various dimensions revisited. Int. J. Mod. Phys. **A13**, 431–490 (1998). (hep-th/9612105)
2. L. Andrianopoli, R. D'Auria, E. Orazi, M. Trigiante, First order description of black holes in moduli space. J. High Energy Phys. **0711**, 032 (2007). (arXiv:0706.0712 (hep-th))
3. L. Andrianopoli, R. D'Auria, S. Ferrara, M. Trigiante, Fake superpotential for large and small extremal black holes. J. High Energy Phys. **1008**, 126 (2010). (arXiv:1002.4340 (hep-th))
4. L. Andrianopoli, R. D'Auria, E. Orazi, M. Trigiante, First order description of D = 4 static black holes and the Hamilton-Jacobi equation. Nucl. Phys. **B833**, 1–16 (2010). (arXiv:0905.3938 (hep-th))
5. V. Balasubramanian, J. de Boer, S. El-Showk, I. Messamah, Black holes as effective geometries. Class. Quantum Gravity **25**, 214004 (2008). (arXiv:0811.0263 (hep-th))
6. K. Behrndt, R. Kallosh, J. Rahmfeld, M. Shmakova, W.K. Wong, STU black holes and string triality. Phys. Rev. **D54**, 6293–6301 (1996). (arXiv:hep-th/9608059 (hep-th))
7. J.D. Bekenstein, Black holes and entropy. Phys. Rev. **D7**, 2333–2346 (1973)

8. S. Bellucci, S. Ferrara, M. Gunaydin, A. Marrani, Charge orbits of symmetric special geometries and attractors. Int. J. Mod. Phys. **A21**, 5043–5098 (2006). (hep-th/0606209)
9. S. Bellucci, S. Ferrara, A. Marrani, A. Yeranyan, $d = 4$ black hole attractors in $N = 2$ supergravity with Fayet-Iliopoulos terms. Phys. Rev. **D77**, 085027 (2008). (arXiv:0802.0141 (hep-th))
10. S. Bellucci, S. Ferrara, A. Shcherbakov, A. Yeranyan, Black hole entropy, flat directions and higher derivatives. J. High Energy Phys. **0910**, 024 (2009). (arXiv:0906.4910 (hep-th))
11. S. Bellucci, S. Ferrara, A. Marrani, A. Yeranyan, stu black holes unveiled. (arXiv:0807.3503 (hep-th))
12. I. Bena, N.P. Warner, One ring to rule them all ... and in the darkness bind them? Adv. Theor. Math. Phys. **9**, 667–701 (2005). (hep-th/0408106)
13. I. Bena, N.P. Warner, Black holes, black rings and their microstates. Lect. Notes Phys. **755**, 1–92 (2008). (hep-th/0701216)
14. I. Bena, G. Dall'Agata, S. Giusto, C. Ruef, N.P. Warner, Non-BPS black rings and black holes in Taub-NUT. J. High Energy Phys. **0906**, 015 (2009). (arXiv:0902.4526 (hep-th))
15. I. Bena, S. Giusto, C. Ruef, N.P. Warner, Multi-center non-BPS black holes: the solution. J. High Energy Phys. **0911**, 032 (2009). (arXiv:0908.2121 (hep-th))
16. I. Bena, S. Giusto, C. Ruef, N.P. Warner, A (running) bolt for new reasons. J. High Energy Phys. **0911**, 089 (2009). (arXiv:0909.2559 (hep-th))
17. I. Bena, S. Giusto, C. Ruef, N.P. Warner, Supergravity solutions from floating branes. J. High Energy Phys. **1003**, 047 (2010). (arXiv:0910.1860 (hep-th))
18. I. Bena, N. Bobev, S. Giusto, C. Ruef, N.P. Warner, An infinite-dimensional family of black-hole microstate geometries. J. High Energy Phys. **1103**, 022 (2011). (arXiv:1006.3497 (hep-th))
19. G. Bossard, Y. Michel, B. Pioline, Extremal black holes, nilpotent orbits and the true fake superpotential. J. High Energy Phys. **1001**, 038 (2010). (arXiv:0908.1742 (hep-th))
20. I. Bredberg, C. Keeler, V. Lysov, A. Strominger, Cargese lectures on the Kerr/CFT correspondence. (arXiv:1103.2355 (hep-th))
21. S.L. Cacciatori, D. Klemm, Supersymmetric AdS(4) black holes and attractors. J. High Energy Phys. **1001**, 085 (2010). (arXiv:0911.4926 (hep-th))
22. A. Celi, A. Ceresole, G. Dall'Agata, A. Van Proeyen, M. Zagermann, On the fakeness of fake supergravity. Phys. Rev. **D71**, 045009 (2005). (hep-th/0410126)
23. A. Ceresole, G. Dall'Agata, Flow equations for Non-BPS extremal black holes. J. High Energy Phys. **0703**, 110 (2007). (hep-th/0702088)
24. A. Ceresole, R. D'Auria, S. Ferrara, The symplectic structure of $N = 2$ supergravity and its central extension. Nucl. Phys. Proc. Suppl. **46**, 67–74 (1996). (hep-th/9509160)
25. A. Ceresole, G. Dall'Agata, S. Ferrara, A. Yeranyan, First order flows for $N = 2$ extremal black holes and duality invariants. Nucl. Phys. **B824**, 239–253 (2010). (arXiv:0908.1110 (hep-th))
26. A. Ceresole, S. Ferrara, A. Marrani, Small $N = 2$ extremal black holes in special geometry. Phys. Lett. **B693**, 366–372 (2010). (arXiv:1006.2007 (hep-th))
27. A. Ceresole, G. Dall'Agata, S. Ferrara, A. Yeranyan, Universality of the superpotential for $d = 4$ extremal black holes. Nucl. Phys. **B832**, 358–381. (arXiv:0910.2697 (hep-th))
28. M. Cvetic, C.M. Hull, Black holes and U duality. Nucl. Phys. **B480**, 296–316 (1996). (hep-th/9606193)
29. M. Cvetic, A.A. Tseytlin, Phys. Rev. **D 53**, 5619 (1996) (Erratum-ibid. D 55, 3907 (1997)) (hep-th/9512031)
30. M. Cvetic, D. Youm, Phys. Rev. **D 53**, 584 (1996) (hep-th/9507090)
31. G. Dall'Agata, A. Gnecchi, Flow equations and attractors for black holes in $N = 2$ U(1) gauged supergravity. J. High Energy Phys. **1103**, 037 (2011). (arXiv:1012.3756 (hep-th))
32. G. Dall'Agata, S. Giusto, C. Ruef, U-duality and non-BPS solutions. J. High Energy Phys. **1102**, 074 (2011). (arXiv:1012.4803 (hep-th))
33. J.R. David, G. Mandal, S.R. Wadia, Microscopic formulation of black holes in string theory. Phys. Rept. **369**, 549–686 (2002). (hep-th/0203048)

34. F. Denef, Supergravity flows and D-brane stability. J. High Energy Phys. **0008**, 050 (2000). (hep-th/0005049)
35. F. Denef, G.W. Moore, Split states, entropy enigmas, holes and halos. (hep-th/0702146 (HEP-TH))
36. S. Ferrara, M. Gunaydin, Orbits of exceptional groups, duality and BPS states in string theory. Int. J. Mod. Phys. **A13**, 2075–2088 (1998). (hep-th/9708025)
37. S. Ferrara, R. Kallosh, Supersymmetry and attractors. Phys. Rev. **D54**, 1514–1524 (1996). (hep-th/9602136)
38. S. Ferrara, R. Kallosh, Universality of supersymmetric attractors. Phys. Rev. **D54**, 1525–1534 (1996). (hep-th/9603090)
39. S. Ferrara, A. Marrani, On the moduli space of non-BPS attractors for $N = 2$ symmetric manifolds. Phys. Lett. **B652**, 111–117 (2007). (arXiv:0706.1667 (hep-th))
40. S. Ferrara, R. Kallosh, A. Strominger, $N = 2$ extremal black holes. Phys. Rev. **D52**, 5412–5416 (1995). (hep-th/9508072)
41. S. Ferrara, G.W. Gibbons, R. Kallosh, Black holes and critical points in moduli space. Nucl. Phys. **B500**, 75–93 (1997). (hep-th/9702103)
42. S. Ferrara, J.M. Maldacena, Branes, central charges and U duality invariant BPS conditions. Class. Quantum Gravity **15**, 749–758 (1998). (hep-th/9706097)
43. S. Ferrara, A. Gnecchi, A. Marrani, $d = 4$ Attractors, effective horizon radius and fake supergravity. Phys. Rev. **D78**, 065003 (2008). (arXiv:0806.3196 (hep-th))
44. S. Ferrara, A. Marrani, Matrix norms, BPS bounds and marginal stability in $N = 8$ supergravity. J. High Energy Phys. **1012**, 038 (2010). (arXiv:1009.3251 (hep-th))
45. S. Ferrara, A. Marrani, E. Orazi, Split attractor flow in $N = 2$ minimally coupled supergravity. Nucl. Phys. **B846**, 512–541 (2011). (arXiv:1010.2280 (hep-th))
46. P. Fre, A.S. Sorin, M. Trigiante, Integrability of supergravity black holes and new tensor classifiers of regular and nilpotent orbits. (arXiv:1103.0848 (hep-th)). W. Chemissany, P. Fre, J. Rosseel, A.S. Sorin, M. Trigiante, T. Van Riet, Black holes in supergravity and integrability. JHEP **1009**, 80 (2010). (arXiv:1007.3209 (hep-th))
47. D.Z. Freedman, C. Nunez, M. Schnabl, K. Skenderis, Fake supergravity and domain wall stability. Phys. Rev. **D69**, 104027 (2004). (hep-th/0312055)
48. M.K. Gaillard, B. Zumino, Duality rotations for interacting fields. Nucl. Phys. **B193**, 221 (1981)
49. D. Gaiotto, W. Li, M. Padi, Non-supersymmetric attractor flow in symmetric spaces. J. High Energy Phys. **0712**, 093 (2007). (arXiv:0710.1638 (hep-th))
50. P. Galli, K. Goldstein, S. Katmadas, J. Perz, First-order flows and stabilisation equations for non-BPS extremal black holes. (arXiv:1012.4020 (hep-th))
51. P. Galli, T. Ortin, J. Perz, C.S. Shahbazi, Non-extremal black holes of $N = 2$, $d = 4$ supergravity. (arXiv:1105.3311 (hep-th))
52. A.M. Ghez, S. Salim, N.N. Weinberg, J.R. Lu, T. Do, J.K. Dunn, K. Matthews, M. Morris et al., Measuring distance and properties of the Milky Way's central supermassive black hole with stellar orbits. Astrophys. J. **689**, 1044–1062 (2008). (arXiv:0808.2870 (astro-ph))
53. G.W. Gibbons, R. Kallosh, B. Kol, Moduli, scalar charges, and the first law of black hole thermodynamics. Phys. Rev. Lett. **77**, 4992–4995 (1996). (hep-th/9607108)
54. E.G. Gimon, F. Larsen, J. Simon, Black holes in supergravity: the Non-BPS branch. J. High Energy Phys. **0801**, 040 (2008). (arXiv:0710.4967 (hep-th))
55. E.G. Gimon, F. Larsen, J. Simon, Constituent model of extremal non-BPS black holes. J. High Energy Phys. **0907**, 052 (2009). (arXiv:0903.0719 (hep-th))
56. K. Goldstein, S. Katmadas, Almost BPS black holes. J. High Energy Phys. **0905**, 058 (2009). (arXiv:0812.4183 (hep-th))
57. K. Goldstein, R.P. Jena, G. Mandal, S.P. Trivedi, A C-function for non-supersymmetric attractors. J. High Energy Phys. **0602**, 053 (2006). (hep-th/0512138)
58. S.W. Hawking, Gravitational radiation from colliding black holes. Phys. Rev. Lett. **26**, 1344–1346 (1971)

59. K. Hristov, S. Vandoren, Static supersymmetric black holes in AdS_4 with spherical symmetry. J. High Energy Phys. **1104**, 047 (2011). (arXiv:1012.4314 (hep-th))
60. C.M. Hull, P.K. Townsend, Unity of superstring dualities. Nucl. Phys. **B438**, 109–137 (1995). (hep-th/9410167)
61. R. Kallosh, T. Ortin, Phys. Rev. **D 48**, 742 (1993) (hep-th/9302109)
62. R. Kallosh, Multivalued entropy of supersymmetric black holes. J. High Energy Phys. **0001**, 001 (2000). (hep-th/9912053)
63. R. Kallosh, B. Kol, E(7) symmetric area of the black hole horizon. Phys. Rev. **D53**, 5344–5348 (1996). (hep-th/9602014)
64. R. Kallosh, A.D. Linde, M. Shmakova, Supersymmetric multiple basin attractors. J. High Energy Phys. **9911**, 010 (1999). (hep-th/9910021)
65. I. Kanitscheider, K. Skenderis, M. Taylor, Holographic anatomy of fuzzballs. J. High Energy Phys. **0704**, 023 (2007). (hep-th/0611171)
66. G. Lopes Cardoso, A. Ceresole, G. Dall'Agata, J.M. Oberreuter, J. Perz, First-order flow equations for extremal black holes in very special geometry. J. High Energy Phys. **0710**, 063 (2007). (arXiv:0706.3373 (hep-th))
67. J.M. Maldacena, Black holes in string theory (hep-th/9607235)
68. S.D. Mathur, The Fuzzball proposal for black holes: an elementary review. Fortschr. Phys. **53**, 793–827 (2005). (hep-th/0502050)
69. G.W. Moore, Arithmetic and attractors. (hep-th/9807087)
70. G. Moore, PiTP lectures on BPS states and wall-crossing in d = 4, N = 2 theories, http://www. physics.rutgers.edu/~gmoore/PiTP_July26_2010.pdf
71. T. Ortin, Phys. Rev. **D 47**, 313 (1993)
72. J. Perz, P. Smyth, T. Van Riet, B. Vercnocke, First-order flow equations for extremal and non-extremal black holes. J. High Energy Phys. **0903**, 150 (2009). (arXiv:0810.1528 (hep-th))
73. B. Pioline, Lectures on black holes, topological strings and quantum attractors. Class. Quantum Gravity **23**, S981 (2006). (hep-th/0607227)
74. A. Sen, Nucl. Phys. **B 440, 421** (1995) (hep-th/9411187)
75. K. Skenderis, P.K. Townsend, Gravitational stability and renormalization group flow. Phys. Lett. **B468**, 46–51 (1999). (hep-th/9909070)
76. A. Strominger, Macroscopic entropy of N = 2 extremal black holes. Phys. Lett. **B383**, 39–43 (1996). (hep-th/9602111)
77. A. Strominger, Black holes – the harmonic oscillators of the 21st Century, http://media.physics. harvard.edu/video/?id=COLLOQ_STROMINGER_091310
78. A. Strominger, C. Vafa, Microscopic origin of the Bekenstein-Hawking entropy. Phys. Lett. **B379**, 99–104 (1996). (hep-th/9601029)
79. P.K. Townsend, Black holes: lecture notes. (gr-qc/9707012)
80. R.M. Wald, The 'Nernst theorem' and black hole thermodynamics. Phys. Rev. **D56**, 6467–6474 (1997). (gr-qc/9704008)
81. M. Wijnholt, S. Zhukov, On the uniqueness of black hole attractors. (hep-th/9912002)

Chapter 2
Intersecting Attractors

Jose Francisco Morales

2.1 Introduction

The attractor mechanism [1–4], initially discovered in the context of $\mathcal{N} = 2$ black holes, has been recognized as a universal phenomenon governing any extremal flow in supergravity, i.e. a flow with an AdS horizon. It applies to both BPS and non-BPS black hole solutions in Einstein supergravities [5, 6], ungauged [7] and gauged [8] supergravities with higher derivative interactions and general intersections of brane solutions [9].

In these lectures we review a unifying framework [9, 10] for the attractor mechanism underlying any extremal flow in supergravity. More precisely, we consider general black p-brane solutions built out of intersections of branes (and or fluxes) with AdS_{p+2} near horizon geometry. In complete analogy with what happens in the case of extremal black holes $p = 0$, the solutions can be thought as scalar attractor flows from infinity to a horizon where active scalars becomes fixed to particular values determined entirely in terms of the black p-brane charges. Moreover the near horizon geometry encodes the thermodynamical content (entropy or central charges) of the boundary theory describing the microscopic degrees of freedom of the black p-brane. The scalar flow generalizes the more familiar black hole attractor mechanism to the case where a general set of branes charged under forms of various ranks intersect on an extended p-dimensional hyperplane.

We focus on static, asymptotically flat, spherically symmetric and extremal black p-brane solutions in supergravities at the two derivative level. The analysis combines standard attractor techniques based on the extremization of the black hole central charge [1–4] and the so-called "entropy function formalism" introduced in [7] (see [11, 12] for reviews and complete lists of references). Like for black

J.F. Morales (✉)
INFN, Section of Rome Tor Vergata, Rome, Italy
e-mail: morales@roma2.infn.it

S. Bellucci (ed.), *Supersymmetric Gravity and Black Holes*, Springer Proceedings in Physics 142, DOI 10.1007/978-3-642-31380-6_2,

holes carrying vector-like charges, we define the entropy function for black p-branes as the Legendre transform with respect to the brane charges of the supergravity action evaluated at the near-horizon geometry. The resulting entropy function can be written as a sum of a gravitational term and an effective potential V_{eff} given as a superposition of the kinetic energies of the forms under which the brane is charged. Extremization of this effective potential gives rise to the attractor equations which determine the values of the scalars at the horizon as functions of the brane charges. In particular, the entropy function itself can be expressed in terms of the U-duality invariants built from these charges and it is proportional to the central charge of the dual CFT_{p+1} living on the AdS_{p+2} boundary. The attractor flow can then be thought of as a c-flow towards the minimum of the *supergravity c-function* [13, 14]. Interestingly, the central charges for extremal black p-branes satisfy an area law formula generalizing the famous Bekenstein-Hawking result for black holes.
We divide the review into two parts:

- In the first part, we illustrate the entropic attractor algorithm in the context of extremal black holes and black strings in $\mathcal{N} = (1, 1)$ supergravity in $D = 6$ dimensions (see [9] for similar results in $D = 7, 8$). We derive the entropy function F and the near-horizon geometry via extremization of F. At the extremum, the entropy function results into a U-duality invariant combination of the brane charges reproducing the black hole entropy and the black string central charge, respectively. Scalars fall into two classes: "fixed scalars" with strictly positive masses and "flat scalars" not fixed by the attractor equations. Flat scalars span the moduli space of the solution. The moduli spaces will be given by symmetric product spaces that can be interpreted as the intersection of the charge orbits of the various branes entering in the solution. In addition one finds extra "geometric moduli" (radii and Wilson lines) that are not fixed by the attractors. Entropy and central charges are entirely determined in terms of the black hole(string) charges and do not depend on the moduli of the solution.
- In the second part we apply the entropic formalism to the study of AdS_4 flux vacua in $\mathcal{N} = 2$ gauged supergravities with an arbitrary number of vector and hyper-multiplets [10]. We find generically a tower of AdS vacua with $\mathcal{N} = 1$ unbroken supersymmetry and two $\mathcal{N} = 0$ towers. We show that both supersymmetric and non-supersymmetric solutions solve a supersymmetry inspired system of linear differential equations. Finally we derive a U-duality invariants formula for the cosmological constant characterizing the vacuum solution.

2.2   Area Law for Central Charges

Before specifying to a particular supergravity theory, here we derive a universal Bekenstein-Hawking like formula underlying any gravity flow (supersymmetric or not) ending on an AdS point. Let $AdS_d \times \Sigma_m$, with Σ_m a product of Einstein spaces, be the near-horizon geometry of an extremal black $(d - 2)$-brane solution in

$D = d + m$ dimensions. After reduction along Σ_m this solution can be thought as the vacuum of a gauged gravity theory in d dimensions. To keep the discussion, as general as possible, we analyze the solution from its d-dimensional perspective. The only fields that can be turned on consistently with the AdS_d symmetries are constant scalar fields. Therefore we can describe the near-horizon dynamics in terms of a gravity theory coupled to scalars φ^i with a potential V_d. The potential V_d depends on the details of the higher-dimensional theory. The "entropy function" is given by evaluating this action at the AdS_d near horizon geometry (with constant scalars $\varphi^i \approx u^i$)

$$F = -\frac{1}{16\pi G_d} \int d^d x \sqrt{-g}\, (R - V_d) = \frac{\Omega_{AdS_d}\, r^d_{AdS}}{16\pi G_d} \left\{ \frac{d(d-1)}{r^2_{AdS}} + V_d \right\}, \quad (2.1)$$

with r_{AdS} the AdS radius and Ω_{AdS_d} the regularized volume of an AdS slice of radius one. Following [15] we take for Ω_{AdS_d} the finite part of the AdS volume integral when the cut off is sent to infinity. More precisely we write the AdS metric

$$ds^2 = r^d_{AdS}(d\rho^2 - \sinh^2\!\rho\, d\tau^2 + \cosh^2\!\rho\, d\Omega^2_{d-2}), \quad (2.2)$$

with $\tau \in [0, 2\pi]$, $0 \leq \rho \leq \cosh^{-1} r_0$ and $d\Omega_{d-2}$ the volume form of a unitary (d-2)-dimensional sphere. The regularized volume Ω_{AdS_d} is then defined as the (absolute value of the) finite part of the volume integral $\int d^d x \sqrt{-g}$ in the limit $r_0 \to \infty$. This results into

$$\Omega_{AdS_d} = \frac{2\pi}{(d-1)} \Omega_{d-2}. \quad (2.3)$$

A different prescription for the volume regularization leads to a redefinition of the entropy function by a charge independent irrelevant constant. The "entropy" and near-horizon geometry follow from the extremization of the entropy function F with respect to the fixed scalars u^i and the radius r_{AdS}

$$\frac{\partial F}{\partial u^i} \propto \frac{\partial V_d}{\partial u^i} \overset{!}{\equiv} 0,$$

$$\frac{\partial F}{\partial r_{AdS}} \propto r^2_{AdS}\, V_d + (d-1)(d-2) \overset{!}{\equiv} 0. \quad (2.4)$$

The first equation determines the values of the scalars at the horizon. The second equation determines the radius of AdS in terms of the value of the potential at the minimum. Notice that solutions exist only if the potential V_d is negative. Indeed, as we will see in the next section, V_d is always composed from a part proportional to a positive definite effective potential V_{eff} generated by the higher dimensional brane charges and a negative contribution $-R_\Sigma$ related to the constant curvature of the internal space Σ (see Eq. 2.30 below). The "entropy" is given by evaluating F at the extremum and can be written in the suggestive form

$$F = \frac{\Omega_{d-2} \, r_{\text{AdS}}^{d-2}}{4 \, G_d} = \Omega_{d-2} \, r_{\text{AdS}}^{d-2} \, \frac{A}{4 \, G_D}, \qquad (2.5)$$

where A denotes the area of Σ_m, Ω_{d-2} is the volume of the unit (d-2)-sphere, and $G_D = A G_d$ the D-dimensional Newton constant. For black holes ($d = 2$), this formula is nothing than the well known Bekenstein-Hawking entropy formula $S = \frac{A}{4 G_D}$ and it shows that F can be identified with the black hole entropy. For black strings ($d = 3$), $\frac{3}{\pi} F = \frac{3 r_{\text{AdS}}}{2 G_3}$ reproduces the central charge c of the two-dimensional CFT living on the AdS_3 boundary [16]. In general, the scaling of (2.5) with the AdS radius matches that of the supergravity c-$function$ introduced in [13] and it suggests that F can be interpreted as the critical value of the central charge c reached at the end of the attractor flow.

2.3   The Entropy Function

The bosonic action of supergravity in D-dimensions can be written as

$$S_{\text{SUGRA}} = \int \left(R * \mathbb{1} - \tfrac{1}{2} g_{ij}(\phi) \, d\phi^i \wedge *d\phi^j - \tfrac{1}{2} N_{\Lambda_n \Sigma_n}(\phi^i) \, F_n^{\Lambda_n} \wedge *F_n^{\Sigma_n} + \mathcal{L}_{\text{WZ}} \right),$$

$$(2.6)$$

with $F_n^{\Lambda_n}$, denoting a set of n-form field strengths, ϕ^i the scalar fields living on a manifold with metric $g_{ij}(\phi)$ and $\mathcal{L}_{\text{WZ}}$ some Wess-Zumino type couplings. The scalar-dependent positive definite matrix $N_{\Lambda_n \Sigma_n}(\phi^i)$ provides the metric for the kinetic term of the n-forms. The sum over n is understood. In the following we will omit the subscript n keeping in mind that both the rank of the forms and the range of the indices Λ depends on n. We will work in units where $16 \pi G_D = 1$, and restore at the end the dependence on G_D. For simplicity we will restrict ourselves here to solutions with trivial Wess-Zumino contributions and this term will be discarded in the following.

We look for extremal black p-brane intersections with near-horizon geometry of topology $M_D = AdS_{p+2} \times S^m \times T^q$. Explicitly we look for solutions with near-horizon geometry

$$ds^2 = r_{\text{AdS}}^2 \, ds_{\text{AdS}_{p+2}}^2 + r_S^2 \, ds_{S^m}^2 + \sum_{k=1}^{q} r_k^2 \, d\theta_k^2,$$

$$F^\Lambda = p_a^\Lambda \alpha^a + e^{\Lambda r} \beta_r, \qquad \phi^i = u^i, \qquad (2.7)$$

with $\mathbf{r} = (r_{\text{AdS}}, r_S, r_k)$, describing the AdS and sphere radii, and u^i denoting the fixed values of the scalar fields at the horizon. α^a and β_r denote the volume forms of the compact $\{\Sigma^a\}$ and non-compact $\{\Sigma_r\}$ cycles, respectively, in M_D. The forms are normalized such as

$$\int_{\Sigma^a} \alpha^b = \delta_a^b, \qquad \int_{\Sigma_r} \beta_s = \delta_s^r. \tag{2.8}$$

They define the volume dependent functions C^{ab}, C_{rs}

$$\int_{M_D} \alpha^a \wedge *\alpha^b = C^{ab}, \qquad \int_{M_D} \beta_r \wedge *\beta_s = C_{rs}, \tag{2.9}$$

describing the cycle intersections. In particular, for the factorized products of AdS space and spheres we consider here, these functions are diagonal matrices with entries

$$C^{ab} = \delta^{ab} \frac{v_D}{\text{vol}(\Sigma^a)^2}, \qquad C_{rs} = \delta_{rs} \frac{v_D}{\text{vol}(\Sigma^r)^2}, \tag{2.10}$$

with v_D the volume of M_D. Integrals over AdS spaces are cut off to a finite volume, according to the discussion around (2.3).

The solutions will be labeled by their electric q_{Ir} and magnetic charges p_a^I defined as

$$p_a^\Lambda = \int_{\Sigma^a} F^\Lambda,$$

$$q_{\Lambda r} = \int_{*\Sigma^r} N_{\Lambda \Sigma} * F^\Sigma = C_{rs} N_{\Lambda \Sigma} e^{\Sigma s}, \tag{2.11}$$

where we denote by $*\Sigma^r$ the complementary cycle to Σ^r in M_D.

Let us now consider the "entropy function" associated to a black p-brane solution with near-horizon geometry (2.7). The entropy function F is defined as the Legendre transform in the electric charges $q_{\Lambda r}$ of S_{SUGRA} evaluated at the near-horizon geometry

$$F = e^{\Lambda r} q_{\Lambda r} - S_{\text{SUGRA}}$$
$$= e^{\Lambda r} q_{\Lambda r} - R v_D + \tfrac{1}{2} N_{\Lambda \Sigma} \, p_a^\Lambda p_b^\Sigma C^{ab} - \tfrac{1}{2} N_{\Lambda \Sigma} e^{\Lambda r} e^{\Sigma s} C_{rs}, \tag{2.12}$$

The fixed values of r, u^i, e^{Ir} at the horizon can be found via extremization of F with respect to r, u^i, and e^{Ir}:

$$\frac{\partial F}{\partial r} = \frac{\partial F}{\partial u^i} = \frac{\partial F}{\partial e^{\Lambda r}} = 0. \tag{2.13}$$

From the last equation one finds that

$$q_{\Lambda r} = N_{\Lambda \Sigma} e^{\Sigma s} C_{rs}, \tag{2.14}$$

in agreement with the definition of electric charges (2.11). Solving this set of equations for $e^{\Lambda r}$ in favor of $q_{\Lambda r}$ one finds

$$F(Q,r,u^i) = -R(r)\,v_D(r) + \tfrac{1}{2}Q^T \cdot M(r,u^i) \cdot Q,\qquad(2.15)$$

with

$$M(r,u^i) = \begin{pmatrix} N_{\Lambda\Sigma}(u^i)C^{ab}(r) & 0 \\ 0 & N^{\Lambda\Sigma}(u^i)C^{rs}(r) \end{pmatrix},\qquad Q\begin{pmatrix} p_a^\Sigma \\ q_{\Sigma r} \end{pmatrix},\qquad(2.16)$$

and $N^{\Lambda\Sigma}$, C^{rs} denoting the inverse of $N_{\Lambda\Sigma}$ and C_{rs} respectively.

It is convenient to introduce the scalar and form intersection "vielbeine" $\mathcal{V}_\Lambda^M$, J^{ab}, J'^{rs} according to

$$N_{\Lambda\Sigma} = \mathcal{V}_\Lambda^M \mathcal{V}_\Sigma^N \delta_{MN}, \qquad C^{ab} = J^{ac}J^{bc}, \qquad C^{rs} = J'^{rt}J'^{st}.\qquad(2.17)$$

From (2.10) one finds for the factorized products of AdS space and spheres

$$J^{ab} = \delta^{ab}\frac{v_D^{1/2}}{\mathrm{vol}(\Sigma^a)}, \qquad J'^{rs} = \delta^{rs}\frac{\mathrm{vol}(\Sigma^r)}{v_D^{1/2}}.\qquad(2.18)$$

The electric and magnetic central charges can be written in terms of these quantities as

$$Z_{\mathrm{mag}}^{Ma} = \mathcal{V}_\Lambda^M J^{ba} p_b^\Lambda, \qquad Z_{\mathrm{el},M}^r = (\mathcal{V}^{-1})_M^\Lambda J'^{sr} q_{\Lambda s}.\qquad(2.19)$$

Combining (2.17) and (2.19) one can rewrite the scalar dependent part of the entropy function as the effective potential

$$V_{\mathrm{eff}} = \tfrac{1}{2}Q^T \cdot M(r,u^i) \cdot Q = \tfrac{1}{2}Z_{\mathrm{mag}}^{Ma}Z_{\mathrm{mag}}^{Ma} + \tfrac{1}{2}Z_{\mathrm{el},M}^r Z_{\mathrm{el},M}^r.\qquad(2.20)$$

For the $n = D/2$-forms in even dimensions the argument is similar, except for the possibility of an additional topological term

$$S_{\mathrm{SUGRA}} = \int \left(R*\mathbb{1} - \tfrac{1}{2}\mathcal{I}_{\Lambda\Sigma}(\phi^i)\, F_n^\Lambda \wedge *F_n^\Sigma - \tfrac{1}{2}\mathcal{R}_{\Lambda\Sigma}(\phi^i)\, F_n^\Lambda \wedge F_n^\Sigma \right),\qquad(2.21)$$

(note that $\mathcal{R}_{\Lambda\Sigma} = \epsilon\mathcal{R}_{\Sigma\Lambda}$, with $\epsilon = (-1)^{[D/2]}$). Following the same steps as before one finds

$$V_{\mathrm{eff}}\tfrac{1}{2}Q^T \cdot M(r,u^i) \cdot Q,\qquad(2.22)$$

with

$$M(r,u^i) \equiv C^{ab}\begin{pmatrix} (\mathcal{I} + \epsilon\mathcal{R}\mathcal{I}^{-1}\mathcal{R})_{\Lambda\Sigma} & \epsilon(\mathcal{R}\mathcal{I}^{-1})_\Lambda^\Sigma \\ (\mathcal{I}^{-1}\mathcal{R})_\Sigma^\Lambda & (\mathcal{I}^{-1})^{\Lambda\Sigma} \end{pmatrix},\qquad Q\begin{pmatrix} p_a^\Lambda \\ q_{\Lambda a} \end{pmatrix}.\qquad(2.23)$$

For $\mathcal{R} = 0$ we are back to the diagonal matrix (2.16). In general, thus we obtain for the $D/2$-forms an effective potential

$$V_{\text{eff}} = \tfrac{1}{2} \, Q^T \cdot M(r, u^i) \cdot Q = \tfrac{1}{2} \, Z^{Ma} \, Z^{Ma}, \qquad (2.24)$$

with

$$Z^{Ma} = J^{ba} \, (\mathcal{V}_\Lambda^M \, p_b^\Lambda + \mathcal{V}^{\Lambda M} \, q_{\Lambda b}), \qquad (2.25)$$

where $\mathcal{V}_I^M = (\mathcal{V}_\Lambda^M, \mathcal{V}^{\Lambda M})$ is the coset representative.

Summarizing, in the case of a general supergravity with bosonic action (2.6) the entropy function is given by

$$F(Q, r, u^i) = -R(r) v_D(r) + V_{\text{eff}}(u^i, r), \qquad (2.26)$$

with the *intersecting-branes effective potential*

$$
\begin{aligned}
V_{\text{eff}} &= \tfrac{1}{2} \sum_n Q_n^T \cdot M_n(r, u^i) \cdot Q_n \\
&= \tfrac{1}{2} Z^{Ma} Z^{Ma} + \tfrac{1}{2} \sum_{n \neq D/2} \left(Z_{\text{mag}}^{M_n a_n} \, Z_{\text{mag}}^{M_n a_n} + Z_{\text{el}, M_n}^{r_n} \, Z_{\text{el}, M_n}^{r_n} \right),
\end{aligned}
\qquad (2.27)
$$

where the first contribution in the second line comes from the $n = D/2$ forms. Notice that there are two types of interference between the potentials coming from forms of different rank: First, they in general depend on a common set of scalar fields and second, they carry a non-trivial dependence on the AdS and the sphere radii. Besides this important difference the critical points of the effective potential can be studied with the standard attractor techniques for vector like charged black holes.

The near-horizon geometry follows from the extremization equations

$$\nabla V_{\text{eff}} \equiv \partial_{u^i} V_{\text{eff}} \, du^i \, \tfrac{1}{2} \sum_n Q_n^T \cdot \nabla M_n(r, u^i) \cdot Q_n \overset{!}{=} 0, \qquad (2.28)$$

$$\partial_r \left[-R(r) \, v_D(r) + V_{\text{eff}}(u^i, r) \right] \overset{!}{=} 0 \,. \qquad (2.29)$$

We conclude this section by noticing that after reduction to AdS_d, the D-dimensional effective potential V_{eff} combines with the contribution coming from the scalar curvature R_Σ of the internal manifold into the d-dimensional scalar potential

$$V_d = \frac{1}{v_D} V_{\text{eff}} - R_\Sigma \qquad (2.30)$$

appearing in (2.1). Notice that the resulting potential is not positive defined and therefore an AdS vacuum is supported.

2.4 $\mathcal{N} = (1, 1)$ in $D = 6$

In this section we illustrate the entropy formalism in the context of $\mathcal{N} = (1, 1)$ in $D = 6$ dimensions. In $D = 6$ dimensions, supergravities contain scalars, vectors and two forms and one finds both black hole and black string extremal solutions.

2.4.1 $\mathcal{N} = (1, 1)$, $D = 6$ Supersymmetry Algebra

The half-maximal $(1, 1)$, $D = 6$ Poincaré supersymmetry algebra has Weyl pseudo-Majorana supercharges and $\mathcal{R}$-symmetry $SO\,(4) \sim SU\,(2)_L \times SU\,(2)_R$. Its central extension reads as follows (see e.g. [17–19])

$$\left\{ \mathcal{Q}^A_\gamma, \mathcal{Q}^B_\delta \right\} = \gamma^\mu_{\gamma\delta} Z^{[AB]}_\mu + \gamma^{\mu\nu\rho}_{\gamma\delta} Z^{(AB)}_{\mu\nu\rho}; \tag{2.31}$$

$$\left\{ \mathcal{Q}^{\dot A}_{\dot\gamma}, \mathcal{Q}^{\dot B}_{\dot\delta} \right\} = \gamma^\mu_{\dot\gamma\dot\delta} Z^{[\dot A \dot B]}_\mu + \gamma^{\mu\nu\rho}_{\dot\gamma\dot\delta} Z^{(\dot A \dot B)}_{\mu\nu\rho}; \tag{2.32}$$

$$\left\{ \mathcal{Q}^A_\gamma, \mathcal{Q}^{\dot A}_{\dot\delta} \right\} = C_{\gamma\dot\delta} Z^{A\dot A} + \gamma^{\mu\nu}_{\gamma\dot\delta} Z^{A\dot A}_{\mu\nu}, \tag{2.33}$$

where $A, \dot A = 1, 2$, so that the (L,R)-chiral supercharges are $SU(2)_{(L,R)}$-doublets.

Notice that $Z^{(AB)}_{\mu\nu\rho} = Z^{(\dot A \dot B)}_{\mu\nu\rho} = 0$, because the presence of the term $Z^{(AB)}_{\mu\nu\rho}$ is inconsistent with the bound $p \leq D - 4$, due to the assumed asymptotical flatness of the (intersecting) black p-brane space-time background.

Strings can be dyonic, and are associated to the central charges $Z^{[AB]}_\mu$, $Z^{[\dot A \dot B]}_\mu$ in the $(\mathbf{1}, \mathbf{1})$ of the $\mathcal{R}$-symmetry group. They are embedded in the $\mathbf{1}_\pm$ (here and below the subscripts denote the weight of $SO(1, 1)$) of the U-duality group $SO\,(1, 1) \times SO\,(4, n_V)$. On the other hand, black holes and their magnetic duals (black two-branes) are associated to $Z^{A\dot A}$, $Z^{A\dot A}_{\mu\nu}$ in the $(\mathbf{2}, \mathbf{2}')$ of $SO\,(4)$, and they are embedded in the $(\mathbf{n}_V + \mathbf{4})_{\pm\frac{1}{2}}$ of $SO\,(1, 1) \times SO\,(4, n_V)$.

In our analysis, the corresponding central charges are denoted respectively by Z_+ and Z_- for dyonic strings, and by $Z_{\mathrm{el}, A\dot A}$ and $Z_{\mathrm{mag}, A\dot A}$ for black holes and their magnetic duals.

2.4.2 $\mathcal{N} = (1, 1)$, $D = 6$ Supergravity

The bosonic field content of half-maximal $\mathcal{N} = (1, 1)$ supergravity in $D = 6$ dimensions coupled to n_V matter (*vector*) multiplets consists of a graviton, $(n_V + 4)$ vector fields with field strengths F_2^M, $M = 1, \ldots, (n_V + 4)$, a three form field strength H_3, and $4n_V + 1$ scalar fields parametrizing the scalar manifold

$$\mathcal{M} = SO(1,1) \times \frac{SO(4,n_V)}{SO(4) \times SO(n_V)}, \qquad \dim_{\mathbb{R}} \mathcal{M} = 4n_V + 1, \qquad (2.34)$$

with the dilaton ϕ spanning $SO(1,1)$, and the $4n_V$ real scalars z^i $(i = 1, \ldots, 4n_V)$ parametrising the quaternionic manifold $\frac{SO(4,n_V)}{SO(4) \times SO(n_V)}$. The U-duality group is $SO(1,1) \times SO(4,n_V)$ and the field strengths transform under this group in the representations

$$F_2^\Lambda : \quad (\mathbf{n}_V + \mathbf{4})_{+\frac{1}{2}},$$

$$H_3 : \quad \mathbf{1}_{\pm 1}. \qquad (2.35)$$

The coset representative $L_\Lambda{}^M$, $\Lambda, M = 1, \ldots, 4 + n_V$, of $\frac{SO(4,n_V)}{SO(4) \times SO(n_V)}$ sits in the $(\mathbf{4}, \mathbf{n}_V)$ representation of the stabilizer $H = SO(4) \times SO(n_V) \sim SU(2)_L \times SU(2)_R \times SO(n_V)$, and satisfies the defining relations

$$L_\Lambda{}^M \eta_{MN} L_\Sigma{}^N = \eta_{\Lambda\Sigma}, \qquad L_\Lambda{}^M \eta^{\Lambda\Sigma} L_\Sigma{}^N = \eta^{MN}, \qquad (2.36)$$

with the $SO(4,n_V)$ metric $\eta_{\Lambda\Sigma}$. It is related to the vielbein $\mathcal{V}_\Lambda^M$ from (2.17) by

$$\mathcal{V}_\Lambda^M = e^{-\phi/2} L_\Lambda{}^M, \qquad (2.37)$$

and its inverse is defined by $L_M^\Lambda L_\Lambda^N = \delta_M^N$. The Maurer-Cartan equations take the form

$$P_{MN} = L_M^\Lambda \, d_z L_{\Lambda N} = L_M^\Lambda \, \partial_i L_{\Lambda N} \, dz^i, \qquad (2.38)$$

where P_{MN} is a symmetric off-diagonal block matrix with non-vanishing entries only in the $(4 \times n_V)$-blocks. Here and below we use δ_{MN} to raise and lower the indices M, N.

The solutions will be specified by the electric and magnetic three-form charges q, p, and the two-form charges p^Λ, q_Σ. The quadratic and cubic U-duality invariants that can be built from these charges are

$$\mathcal{I}_2 = pq, \qquad \mathcal{I}_3 = \tfrac{1}{2}\eta_{\Lambda\Sigma} \, p^\Lambda p^\Sigma p, \qquad \mathcal{I}_3' = \tfrac{1}{2}\eta^{\Lambda\Sigma} \, q_\Lambda q_\Sigma q. \qquad (2.39)$$

The central charges (2.19) and (2.25) are given by

$$Z_{\mathrm{mag},M} = e^{-\phi/2} J_2 L_{\Lambda M} \, p^\Lambda, \qquad Z_{\mathrm{el},M} = e^{\phi/2} J_2' L_M^\Lambda \, q_\Lambda,$$

$$Z_\pm = \frac{1}{\sqrt{2}} J_3 (e^\phi \, p \pm e^{-\phi} \, q). \qquad (2.40)$$

Using (2.36), the U-duality invariants (2.39) can be rewritten in terms of the central charges as

$$\tfrac{1}{2}\left(Z_+^2 - Z_-^2\right) = J_3^2\,\mathcal{I}_2,$$

$$\tfrac{1}{2\sqrt{2}}\eta^{MN}\,Z_{\mathrm{mag},M}Z_{\mathrm{mag},N}\left(Z_+ + Z_-\right) = \left(J_3 J_2^2\right)\mathcal{I}_3,$$

$$\tfrac{1}{2\sqrt{2}}\eta^{MN}\,Z_{\mathrm{el},M}Z_{\mathrm{el},N}\left(Z_+ - Z_-\right) = \left(J_3 J_2'^2\right)\mathcal{I}_3'. \tag{2.41}$$

The effective potential V_{eff} (2.27) for this theory is given by

$$V_{\mathrm{eff}} = \tfrac{1}{2}Z_+^2 + \tfrac{1}{2}Z_-^2 + \tfrac{1}{2}Z_{\mathrm{el},M}^2 + \tfrac{1}{2}Z_{\mathrm{mag},M}^2\,. \tag{2.42}$$

From the Maurer-Cartan equations (2.38) one derives

$$\nabla Z_{\mathrm{mag},M} = -P_{MN}\,Z_{\mathrm{mag},N} - \tfrac{1}{2}P_\phi\,Z_{\mathrm{mag},M}\,,$$

$$\nabla Z_{\mathrm{el},M} = P_{MN}\,Z_{\mathrm{el},N} + \tfrac{1}{2}P_\phi\,Z_{\mathrm{el},M}\,,$$

$$\nabla Z_\pm = P_\phi\,Z_\mp\,. \tag{2.43}$$

with $P_\phi = d\phi$. The attractor equations (2.28) thus translate into

$$P_{MN}\left(Z_{\mathrm{el},M}Z_{\mathrm{el},N} - Z_{\mathrm{mag},M}Z_{\mathrm{mag},N}\right) + P_\phi\left(2Z_+Z_- - \tfrac{1}{2}Z_{\mathrm{mag},M}^2 + \tfrac{1}{2}Z_{\mathrm{el},M}^2\right) \overset{!}{=} 0\,. \tag{2.44}$$

Splitting the index M into $(A\dot{A}) = 1,\ldots,4$, $(A,\dot{A} = 1,2)$ (*central charges sector*) and $I = 5,\ldots,(n_V + 4)$ (*matter charges sector*), and using the fact that only the components $P_{I,A\dot{A}} = P_{A\dot{A},I}$ are non-vanishing, the attractor equations can be written as

$$Z_{\mathrm{el},A\dot{A}}Z_{\mathrm{el},I} - Z_{\mathrm{mag},A\dot{A}}Z_{\mathrm{mag},I} = 0,$$

$$4Z_+Z_- - Z_{\mathrm{mag},A\dot{A}}Z_{\mathrm{mag}}^{A\dot{A}} + Z_{\mathrm{el},A\dot{A}}Z_{\mathrm{el}}^{A\dot{A}} - Z_{\mathrm{mag},I}^2 + Z_{\mathrm{el},I}^2 = 0\,. \tag{2.45}$$

Indices $A, \dot{A}$ are raised and lowered by $\epsilon_{AB}, \epsilon_{\dot{A}\dot{B}}$. We will study the solutions of these equations, their supersymmetry-preserving features, and the corresponding moduli spaces. BPS solutions correspond to the solutions of (2.45) satisfying

$$Z_{\mathrm{mag},I} = Z_{\mathrm{el},I} = 0, \tag{2.46}$$

as follows from the Killing spinor equation $\delta\lambda_A^I \sim T_{\mu\nu}^I \gamma^{\mu\nu}\epsilon_A = 0$ with $T_{\mu\nu}^I$ the matter central charge densities.

Let us finally consider the *moduli space* of the attractor solutions, i.e. the scalar degrees of freedom which are *not* stabilized by the attractor mechanism at the classical level. For homogeneous scalar manifolds this space is spanned by the vanishing eigenvalues of the Hessian matrix $\nabla\nabla V_{\mathrm{eff}}$ at the critical point. Using the Maurer-Cartan equations (2.43) one can write $\nabla\nabla V_{\mathrm{eff}}$ at the critical point as

$$\nabla\nabla V_{\text{eff}} = P_{I,A\dot{A}} P_J^{A\dot{A}} \left(2 Z_{\text{el},I} Z_{\text{el},J} + 2 Z_{\text{mag},I} Z_{\text{mag},J} \right)$$

$$+ P^{I,A\dot{A}I} P^{I,B\dot{B}} \left(2 Z_{\text{el},A\dot{A}} Z_{\text{el},B\dot{B}} + 2 Z_{\text{mag},A\dot{A}} Z_{\text{mag},B\dot{B}} \right)$$

$$+ P_\phi P_\phi \left(2Z_+^2 + 2Z_-^2 + \tfrac{1}{2} Z_{\text{mag},M}^2 + \tfrac{1}{2} Z_{\text{el},M}^2 \right)$$

$$+ 2 P_\phi P^{I,A\dot{A}} \left(Z_{\text{el},I} Z_{\text{el},A\dot{A}} + Z_{\text{mag},I} Z_{\text{mag},A\dot{A}} \right)$$

$$= H_{IA\dot{A},JB\dot{B}} P^{I,A\dot{A}} P^{J,B\dot{B}} + 2 H_{IA\dot{A},\phi} P^{I,A\dot{A}} P_\phi + H_{\phi,\phi} P_\phi P_\phi, \quad (2.47)$$

which defines the Hessian symmetric matrix **H** with components $H_{IA\dot{A},JB\dot{B}}$, $H_{IA\dot{A},\phi}$, $H_{\phi,\phi}$. By explicit evaluation of the Hessian matrix for both BPS and non-BPS solutions we will show that eigenvalues are always zero or positive implying the stability (at the classical level) of the solutions under consideration here. We will now specify to the different near-horizon geometries and study the BPS and non-BPS solutions of the attractor equations.

2.4.3 $AdS_3 \times S^3$

Let us start with an $AdS_3 \times S^3$ near-horizon geometry, in which only the three-form charges (magnetic p and electric q) are switched on (*dyonic black string*). There are no closed two-forms supported by this geometry and therefore two-form charges are not allowed. The near-horizon geometry ansatz can then be written as

$$ds^2 = r_{\text{AdS}}^2 ds_{\text{AdS}_3}^2 + r_S^2 ds_{S^3}^2, \qquad H_3 = p \, \alpha_{S^3} + e \, \beta_{\text{AdS}_3}. \quad (2.48)$$

The attractor equations (2.45) are solved by

$$Z_{\text{mag},M} = Z_{\text{el}M} = Z_- = 0, \quad \text{or equivalently,} \quad (2.49)$$

$$Z_{\text{mag},M} = Z_{\text{el},M} = Z_+ = 0. \quad (2.50)$$

Solution (2.49) has $\mathcal{I}_2 > 0$, whereas solution (2.50) has $\mathcal{I}_2 < 0$; they are both $\frac{1}{4}$-BPS, and they are equivalent, because the considered theory is non-chiral.

Plugging the solutions (2.49) or (2.50) into (2.42) one can write the effective potential at the horizon in the scalar independent form

$$V_{\text{eff}} = \tfrac{1}{2} Z_+^2 + \tfrac{1}{2} Z_-^2 = \tfrac{1}{2} |Z_+^2 - Z_-^2| J_3^2 |\mathcal{I}_2| = \left(\frac{v_{AdS_3}}{v_{S^3}} \right) |\mathcal{I}_2|, \quad (2.51)$$

Extremizing F in r, one finds the entropy function and near-horizon AdS and sphere radii

$$V_{\text{eff}} = \frac{v_{\text{AdS}_3}}{v_{S^3}} |\mathcal{I}_2|, \qquad r_{\text{AdS}} = r_S = \frac{|\mathcal{I}_2|^{1/4}}{2\pi},$$

$$F = v_{\text{AdS}_3 \times S^3} \left(\frac{6}{r_{\text{AdS}}^2} - \frac{6}{r_S^2} \right) + V_{\text{eff}} = |\mathcal{I}_2| . \tag{2.52}$$

Now let us consider the moduli space of the solutions. Plugging (2.49) and (2.50) into (2.47) one finds that the only non-trivial component of the Hessian matrix is

$$H_{\phi\phi} = 2Z_+^2 + 2Z_-^2 = 4V_{\text{eff}} > 0. \tag{2.53}$$

Therefore, the Hessian matrix $\mathbf{H}$ for the $AdS_3 \times S^3$ solution has $4n_V$ vanishing eigenvalues and one strictly positive eigenvalue, corresponding to the dilaton direction. Consequently, the moduli space of non-degenerate attractors with near-horizon geometry $AdS_3 \times S^3$ is the quaternionic symmetric manifold

$$\mathcal{M}_{\text{BPS}} = \frac{SO(4, n_V)}{SO(4) \times SO(n_V)} . \tag{2.54}$$

This result is also evident from the explicit form of the attractor solution $Z_- = 0$: only the dilaton is stabilized, while all other scalars are not fixed since the remaining equations $Z_{\text{el},M} = Z_{\text{mag},M} = 0$ are automatically satisfied for $p^\Lambda = q_\Lambda = 0$.

2.4.4 $AdS_3 \times S^2 \times S^1$

For solutions with near-horizon geometry $AdS_3 \times S^2 \times S^1$, there is no support for electric two-form charges and therefore $e^\Lambda = 0$. We set also the electric three-form charge e to zero otherwise no solutions are found. The near-horizon ansatz becomes

$$ds^2 = r_{\text{AdS}}^2 ds_{\text{AdS}_3}^2 + r_S^2 ds_{S^2}^2 + r_1^2 d\theta^2,$$

$$F_2^\Lambda = p^\Lambda \alpha_{S^2}, \qquad H_3 = p \, \alpha_{S^2 \times S^1}. \tag{2.55}$$

The attractor equations (2.45) admit two types of solutions with non trivial central charges

$$\text{BPS}: \ Z_+ = Z_-, \qquad Z_{\text{mag},A\dot{A}} Z_{\text{mag}}^{A\dot{A}} = 4Z_+^2 ; \tag{2.56}$$

$$\text{non} - \text{BPS}: \ Z_+ = Z_- \qquad Z_{\text{mag},I}^2 = 4Z_+^2. \tag{2.57}$$

Plugging the solution into (2.41) one finds the relation

$$\left| J_2^2 J_3 \mathcal{I}_3 \right| = 2\sqrt{2} \, Z_+^3. \tag{2.58}$$

that allows us to write the effective potential (2.42) at the horizon in the scalar independent form

$$V_{\text{eff}} = 3Z_+^2 = \tfrac{3}{2} \left| J_2^2 J_3 \, \mathcal{I}_3 \right|^{\frac{2}{3}},$$ (2.59)

with

$$(J_2^2 J_3)^{\frac{2}{3}} = \frac{v_{\text{AdS}_3} \, v_{S^1}^{\frac{1}{3}}}{v_{S^2}},$$ (2.60)

The black string central charge and the near-horizon radii follow from r-extremization of the entropy function F and are given by

$$V_{\text{eff}} = \tfrac{3}{2} \frac{v_{\text{AdS}_3}}{v_{S^2}} \, v_{S^1}^{1/3} \left| \mathcal{I}_3 \right|^{\frac{2}{3}}, \qquad r_{\text{AdS}} = 2 r_S = \frac{|\mathcal{I}_3|^{1/3}}{2\pi v_{S^1}^{1/3}},$$

$$F = v_{\text{AdS}_3 \times S^2 \times S^1} \left(\frac{6}{r_{\text{AdS}_3}^2} - \frac{2}{r_{S^2}^2} \right) + V_{\text{eff}} = |\mathcal{I}_3|.$$ (2.61)

Note that the radius of the extra S^1 is not fixed by the extremization equations. Besides this geometric modulus the solutions can be also deformed by turning on Wilson lines for the vector field potentials $A_5^\Lambda = c^\Lambda$. This is in contrast with the more familiar case of black holes in $D = 4, 5$ where the near-horizon geometry is completely fixed at the end of the attractor flow. As we shall see in the following, this will be always the case for extremal black p-branes. "Geometric moduli" describing the shapes ad volumes of the extra circles and constant values of field potentials along these circles remain unfixed at the horizon. It is important to stress, that the extreme value of the entropy function does not depend on the moduli of the solution and is given entirely by a U-duality invariant combination of the black brane charges.

Now, let us consider the moduli spaces of the two solutions. The BPS solution (2.56) has remaining symmetry $SO(3) \times SO(n_V)$, because by using an $SO(4)$ transformation this solution can be recast in the form

$$Z_{\text{mag},A\dot{A}} = 2 z \, \delta_{A1} \delta_{\dot{A}1}, \qquad Z_+ = Z_- = z, \qquad Z_{\text{el},M} = 0.$$ (2.62)

Notice that both choices of sign satisfy the Killing spinor relations (2.46) and therefore correspond to supersymmetric solutions. Plugging (2.62) into the Hessian matrix (2.47) one finds

$$\mathbf{H} = z^2 \begin{pmatrix} 8 \delta_{IJ} \delta_{A1} \delta_{B1} \delta_{\dot{A}1} \delta_{\dot{B}1} & 0_{4n_V \times 1} \\ 0_{1 \times 4n_V} & 6 \end{pmatrix}.$$ (2.63)

This matrix has $3n_V$ vanishing eigenvalues and $n_V + 1$ strictly positive eigenvalues, corresponding to the dilaton direction plus the n_V directions $P_{I,11}$. Consequently, the moduli space of the BPS attractor solution (2.56) with near-horizon geometry $AdS_3 \times S^2 \times S^1$ is the symmetric manifold

$$\mathcal{M}_{\text{BPS}} = \frac{SO(3, n_V)}{SO(3) \times SO(n_V)} . \tag{2.64}$$

More precisely, the scalars along $P_{I,A\dot{A}}$ in the $(\mathbf{4}, \mathbf{n}_V)$ of the group H decompose with respect to the symmetry group $SO(3) \times SO(n_V)$ as:

$$(\mathbf{4}, \mathbf{n}_V) \to (\mathbf{3}, \mathbf{n}_V) \oplus (\mathbf{1}, \mathbf{n}_V) , \tag{2.65}$$

and only the $(\mathbf{1}, \mathbf{n}_V)$ representation is massive, together with the dilaton. The $(\mathbf{3}, \mathbf{n}_V)$ representation remains massless and contains all the massless Hessian modes of the attractor solutions.

The analysis of the moduli space for the non-BPS solution follows closely that for the BPS one. Now the symmetry is $SO(4) \times SO(n_V - 1)$ and using an $SO(n_V)$ transformation such a solution can be recast as follows:

$$Z_{\text{mag},I} = 2z\,\delta_{I1}, \qquad Z_+ = Z_- = z, \qquad Z_{\text{el},M} = Z_{\text{mag},A\dot{A}} = Z_{\text{el},A\dot{A}} = 0. \tag{2.66}$$

Plugging (2.66) into the Hessian matrix (2.47), now one finds

$$\mathbf{H} = z^2 \begin{pmatrix} 8\delta_{A\dot{A}}\delta_{B\dot{B}}\delta_{J1}\delta_{I1} & 0_{4n_V \times 1} \\ & \\ 0_{1 \times 4n_V} & 6 \end{pmatrix} . \tag{2.67}$$

This Hessian matrix has $4(n_V - 1)$ vanishing eigenvalues and $4 + 1$ strictly positive eigenvalues, corresponding to the dilaton direction plus the 4 $P_{1,A\dot{A}}$ directions. Consequently, the *moduli space* of the non-BPS attractor solution with near-horizon geometry $AdS_3 \times S^2 \times S^1$ is the symmetric manifold

$$\mathcal{M}_{\text{nonBPS}} = \frac{SO(4, n_V - 1)}{SO(4) \times SO(n_V - 1)} . \tag{2.68}$$

More precisely, the scalars along $P_{I,A\dot{A}}$ in the $(\mathbf{4}, \mathbf{n}_V)$ of the group H decompose with respect to the symmetry group $SO(4) \times SO(n_V - 1)$ as:

$$(\mathbf{4}, \mathbf{n}_V) \to (\mathbf{4}, \mathbf{n}_V - 1) \oplus (\mathbf{4}, \mathbf{1}) , \tag{2.69}$$

and only the $(\mathbf{4}, \mathbf{1})$ representation is massive, together with the dilaton. The $(\mathbf{4}, \mathbf{n}_V - 1)$ representation remains massless, and it contains all the massless Hessian modes of the attractor solution.

2.4.5 $AdS_2 \times S^3 \times S^1$

For solutions with $AdS_2 \times S^3 \times S^1$ near-horizon geometry, there is no support for magnetic two-form charges and therefore $Z_{\mathrm{mag},M} = 0$. The near-horizon ansatz becomes

$$ds^2 = r_{\mathrm{AdS}}^2 \, ds_{\mathrm{AdS}_2}^2 + r_S^2 \, ds_{S^3}^2 + r_1^2 d\theta^2,$$

$$F_2^\Lambda = e^\Lambda \beta_{\mathrm{AdS}_2}, \qquad H_3 = e \, \beta_{\mathrm{AdS}_2 \times S^1}. \tag{2.70}$$

The fixed-scalar equations (2.44) admit two type of solutions

$$\mathrm{BPS}: \ Z_{\mathrm{mag},M} = Z_{\mathrm{el},I} = 0, \qquad Z_+ = -Z_-, \qquad Z_{\mathrm{el},A\dot{A}} Z_{\mathrm{el}}^{A\dot{A}} = 4Z_+^2,$$

$$\mathrm{non-BPS}: \ Z_{\mathrm{mag},M} = Z_{\mathrm{el},A\dot{A}} = 0, \qquad Z_+ = -Z_-, \qquad Z_{\mathrm{el},I}^2 = 4Z_+^2. \tag{2.71}$$

Now one finds

$$J_2'^2 J_3' \mathcal{I}_3' = 2\sqrt{2} \, Z_+^3, \tag{2.72}$$

and the effective potential (2.42) at the horizon can be written in the scalar independent form

$$V_{\mathrm{eff}} = 3Z_+^2 = \tfrac{3}{2} |J_2'^2 J_3' \mathcal{I}_3'|^{\frac{2}{3}}, \tag{2.73}$$

with

$$(J_2'^2 J_3')^{\frac{2}{3}} = \frac{v_{\mathrm{AdS}_2}}{v_{S^1}^{1/3} v_{S^3}}, \tag{2.74}$$

Extremizing F in the radii r one finds

$$V_{\mathrm{eff}} = \tfrac{3}{2} \frac{v_{\mathrm{AdS}_2}}{v_{S^3} v_{S^1}^{1/3}} |\mathcal{I}_3'|^{\frac{2}{3}}, \qquad r_{\mathrm{AdS}} = \tfrac{1}{2} r_S = \frac{|\mathcal{I}_3'|^{1/6}}{2\pi v_{S^1}^{1/3}},$$

$$F = v_{\mathrm{AdS}_2 \times S^3 \times S^1} \left(\frac{2}{r_{\mathrm{AdS}_2}^2} - \frac{6}{r_{S^3}^2} \right) + V_{\mathrm{eff}} = |\mathcal{I}_3'|^{1/2}. \tag{2.75}$$

for the black hole entropy and AdS and sphere radii. Again, the radius of the extra S^1 is not fixed by the extremization equations. The analysis of the moduli spaces follows *mutatis mutandis* that of the $AdS_3 \times S^2$ attractors (replacing magnetic by electric charges) and the results are again given by the symmetric manifolds (2.64) and (2.68).

Table 2.1 Supersymmetric M-intersections

	0	1	2	3	4	5	6	7	8	9	10	Near-horizon
M2	−	−	•	•	•	•	•	•	•	•	−	$AdS_3 \times S^3 \times T^5$
M5	−	−	•	•	•	•	−	−	−	−	•	
M2	−	•	•	•	•	•	•	•	•	−	−	$AdS_2 \times S^3 \times T^6$
M2	−	•	•	•	•	•	•	−	−	•	•	
M2	−	•	•	•	•	−	−	•	•	•	•	
M5	−	−	•	•	•	•	•	−	−	−	−	$AdS_3 \times S^2 \times T^6$
M5	−	−	•	•	•	−	−	•	•	−	−	
M5	−	−	•	•	•	−	−	−	−	•	•	

2.5 The Lift to Eleven Dimensions

The attractor solutions we have discussed have a simple lift to higher dimensions all the way up to 11-dimensional supergravity. The black string solution with $AdS_3 \times S^3$ near-horizon geometry follow from dimensional reduction of M2M5 branes intersecting on a string. The solution with $AdS_3 \times S^2 \times S^1$ follow from the reduction of a triple M2-intersection on a string. Finally the $AdS_2 \times S^3 \times S^1$ near-horizon geometry corresponds to a triple M5 intersection on a time-like line. The orientations of the M2,M5 branes in the three cases are summarized in Table 2.1.

After dimensional reductions down to $D = 6, 7, 8$-dimensions the solutions expose a variety of charges with respect to forms of various rank. Indeed, a single brane intersection in $D = 11$ leads to different solutions after reduction to D-dimensions depending on the orientation of the M-branes along the internal space. Different solutions carry charges with respect to a different set of forms in the D-dimensional supergravity. They can be fully characterized by U-duality invariants built out of the brane charges. The list of U-duality invariants leading to extremal black p-brane solutions in $D = 6, 7, 8$ dimensions is displayed in Table 2.2. As expected, there is a one-to-one correspondence between the entries in this table and the solutions that follow from extremization of the entropy function in the corresponding D dimensional supergravity. We list also the U-duality groups, the representation content and the corresponding moduli spaces.

2.6 AdS_4 Flux Vacua

AdS$_4$ vacua arise in string theory compactifications with fluxes and can be thought as the near horizon geometries of extremal black brane solutions. In fact, the vacuum conditions of flux compactifications display many analogies with the attractor equations defining the near-horizon limit of extremal black holes. This was first shown in [5] in the context of $\mathcal{N} = 1$ orientifolds of type IIB on Calabi-Yau threefolds. In that paper, first it was observed that the $\mathcal{N} = 1$ scalar potential has a

Table 2.2 Electric and magnetic charges for M-brane intersections. $p = 0, 1$ corresponds to intersections on a black hole and a black string respectively. $q_n(p_n)$ denotes the electric(magnetic) charge of the brane solution and n specifies the rank of the form

d	p	U-duality	Charges	Reps	Moduli space
6	1	$SO(5,5)$	$p_3 q_3$	$\mathbf{10} \times \mathbf{10}$	$\frac{SO(5,4)}{SO(5) \times SO(4)}$
	1		$p_2 p_2 p_3$	$\mathbf{16_s} \times \mathbf{16_s} \times \mathbf{10}$	$\frac{SO(4,3)}{SO(4) \times SO(3)}$
	0		$q_2 q_2 q_3$	$\mathbf{16_c} \times \mathbf{16_c} \times \mathbf{10}$	$\frac{SO(4,3)}{SO(4) \times SO(3)}$
7	1	$SL(5)$	$p_3 q_3$	$\mathbf{5} \times \mathbf{5'}$	$\frac{SL(4)}{SO(4)}$
	1		$p_2 p_2 q_3$	$\mathbf{10'} \times \mathbf{10'} \times \mathbf{5}$	$\frac{SO(3,2)}{SO(3) \times SO(2)}$
	1		$p_3 p_3 p_2$	$\mathbf{5'} \times \mathbf{5'} \times \mathbf{10'}$	$\frac{SL(3)}{SO(3)}$
	0		$q_2 q_2 p_3$	$\mathbf{10} \times \mathbf{10} \times \mathbf{5'}$	$\frac{SO(3,2)}{SO(3) \times SO(2)}$
	0		$q_3 q_3 q_2$	$\mathbf{5} \times \mathbf{5} \times \mathbf{10}$	$\frac{SL(3)}{SO(3)}$
8	1	$SL(3) \times SL(2)$	$p_3 q_3$	$(\mathbf{3'}, \mathbf{1}) \times (\mathbf{3}, \mathbf{1})$	$\left(\frac{SL(2)}{SO(2)} \right)^2$
	1		$p_4 q_4$	$(\mathbf{1}, \mathbf{2}) \times (\mathbf{1}, \mathbf{2})$	$\frac{SL(3)}{SO(3)}$
	1		$p_2 p_3 q_4$	$(\mathbf{3}, \mathbf{2}) \times (\mathbf{3'}, \mathbf{1}) \times (\mathbf{1}, \mathbf{2})$	$\frac{SL(2)}{SO(2)}$
	1		$p_3 p_3 p_3$	$(\mathbf{3'}, \mathbf{1}) \times (\mathbf{3'}, \mathbf{1}) \times (\mathbf{3'}, \mathbf{1})$	$\frac{SL(2)}{SO(2)}$
	0		$q_2 q_3 p_4$	$(\mathbf{3'}, \mathbf{2}) \times (\mathbf{3}, \mathbf{1}) \times (\mathbf{1}, \mathbf{2})$	$\frac{SL(2)}{SO(2)}$
	0		$q_3 q_3 q_3$	$(\mathbf{3}, \mathbf{1}) \times (\mathbf{3}, \mathbf{1}) \times (\mathbf{3}, \mathbf{1})$	$\frac{SL(2)}{SO(2)}$

form close to the $\mathcal{N} = 2$ black hole potential, with the black hole central charge Z replaced by the combination $e^{\frac{K}{2}} W$ involving the $\mathcal{N} = 1$ Kähler potential K and superpotential W, and with the NSNS and RR fluxes playing the role of the black hole electric charges. Then, exploiting the underlying special Kähler geometry of the Calabi-Yau complex structure moduli space, the differential equations both for black hole attractors and for flux vacua were recast in a set of algebraic equations. Explicit attracting Calabi-Yau solutions were derived. A recent review on the subject, including a complete list of references, can be found in [11].

Here we develop the above ideas focussing on a four-dimensional $\mathcal{N} = 2$ setup. We consider general $\mathcal{N} = 2$ gauged supergravities related to type II theories via flux compactification on non-Calabi-Yau manifolds. At the four dimensional level, the family of supergravities we consider can be obtained by deforming the Calabi-Yau effective action. The latter is characterized by the choice of two special Kähler manifolds $\mathcal{M}_1$ and $\mathcal{M}_2$: while $\mathcal{M}_2$ defines the vector multiplet scalar manifold, $\mathcal{M}_1$ determines via c-map the quaternionic manifold $\mathcal{M}_Q$ parameterized by the scalars in the hypermultiplets [20, 21]. The deformation we study is the most general abelian gauging of the Heisenberg algebra of axionic isometries which is always admitted by $\mathcal{M}_Q$ [22–24]. From the point of view of compactifications, this deformation is expected to correspond to a dimensional reduction of type IIA/IIB on $SU(3)$ and $SU(3) \times SU(3)$ structure manifolds (and possibly non-geometric backgrounds) in the presence of general NSNS and RR fluxes [25–30]. In this perspective, our setup provides a unifying framework for the study of $\mathcal{N} = 2$ flux compactifications of type II theories on generalized geometries.

We start our analysis in Sect. 2.6.1 by reconsidering the expression derived in [24] for the $\mathcal{N} = 2$ scalar potential associated with the gauging described above. This expression is manifestly invariant under the symplectic transformations rotating the flux charges together with the symplectic sections of both the special Kähler geometries on $\mathcal{M}_1$ and $\mathcal{M}_2$. We derive a convenient reformulation of the scalar potential in terms of a triplet of Killing prepotentials $\mathcal{P}_x$, $x = 1, 2, 3$, and their covariant derivatives. The $\mathcal{P}_x$ encode the informations about the gauging, and play a role analogous to the central charge Z of $\mathcal{N} = 2$ black holes, or to the covariantly holomorphic superpotential $e^{\frac{K}{2}} W$ of $\mathcal{N} = 1$ compactifications. Interestingly, in our reformulation the only derivatives appearing are the special Kähler covariant derivatives with respect to the coordinates on $\mathcal{M}_1$ and $\mathcal{M}_2$, with no explicit contributions from the $\mathcal{M}_Q$ coordinates orthogonal to $\mathcal{M}_1$. This rewriting is advantageous since it allows to study the hypersector with the powerful techniques of special Kähler geometry.

In Sect. 2.6.2 we move to the study of the extremization equations for the potential. These are spelled out in Sect. 2.6.2, and resemble the attractor equations of $\mathcal{N} = 2$ black holes, though they are more complicated to solve. In Sect. 2.6.3 we present a supersymmetry inspired set of linear differential equations that imply the vacuum equations and encode all (supersymmetric or not) extremal solutions we found. In Sect. 2.6.4 we find explicit AdS solutions for the wide class of supergravities whose scalar manifolds $\mathcal{M}_1$ and $\mathcal{M}_2$ are symmetric with cubic prepotentials. These solutions generalize those derived in [31] in the context of flux compactifications of massive type IIA on coset spaces with $SU(3)$ structure, allowing for an arbitrary number of vector multiplets (Sect. 2.6.4) and hypermultiplets (Sect. 2.6.4), as well as for a large set of fluxes. In Sect. 2.6.5 we derive a manifestly U-invariant expression for the AdS cosmological constant. Appendix A.1 illustrates how the gaugings we consider can arise from dimensional reduction of type II theories on manifolds with $SU(3)$ structure.

2.6.1 Revisiting the $\mathcal{N} = 2$ Flux Potential

In this section we reconsider the scalar potential V derived in [24] by gauging the $\mathcal{N} = 2$ effective action for type II theories on Calabi-Yau three-folds, and we reformulate it in terms of Killing prepotentials $\mathcal{P}_x$ and their special Kähler covariant derivatives. From a dimensional reduction perspective, V arises in general type II flux compactifications on 6d manifolds with $SU(3)$ and $SU(3) \times SU(3)$ structure (and non-geometric backgrounds) preserving eight supercharges [25–30]. An explicit dictionary between the gauging and the fluxes is presented in Appendix A.1 for type IIA on $SU(3)$ structures, together with a further list of references.

Scalars in $\mathcal{N} = 2$ supergravity organize in vector multiplets and hypermultiplets, respectively parameterizing a special Kähler manifold $\mathcal{M}_2$ and a quaternionic manifold $\mathcal{M}_Q$. In the cases of our interest, $\mathcal{M}_Q$ is derived via c-map from a special Kähler submanifold $\mathcal{M}_1$ [20, 21]. For both type IIA/IIB compactifications, we denote by $h_1 + 1$ the number of hypermultiplets (where the 1 is associated with the universal hypermultiplet), and by $h_2 + 1$ the number of vector fields (including the graviphoton in the gravitational multiplet), so that $h_1 = \dim_{\mathbb{C}} \mathcal{M}_1$ and $h_2 = \dim_{\mathbb{C}} \mathcal{M}_2$. Furthermore, we introduce complex coordinates z^i, $i = 1, \ldots h_1$ on $\mathcal{M}_1$, and x^a, $a = 1, \ldots, h_2$ on $\mathcal{M}_2$, and we denote by

$$\Pi_1^{\mathbb{I}} = e^{\frac{K_1}{2}} \begin{pmatrix} Z^I \\ \mathcal{G}_I \end{pmatrix}, \qquad I = (0, i) = 0, 1, \ldots, h_1,$$

$$\Pi_2^{\mathbb{A}} = e^{\frac{K_2}{2}} \begin{pmatrix} X^A \\ \mathcal{F}_A \end{pmatrix}, \qquad A = (0, a) = 0, 1, \ldots, h_2 \qquad (2.76)$$

the covariantly holomorphic symplectic sections of the special Kähler geometry on $\mathcal{M}_1$ and $\mathcal{M}_2$ respectively. All along the paper, indices in a double font like $\mathbb{I}$ and $\mathbb{A}$ correspond to symplectic indices. The respective ranges are $\mathbb{I} = 1, \ldots, 2(h_1 + 1)$ and $\mathbb{A} = 1, \ldots, 2(h_2 + 1)$.

To complete the geometric data on $\mathcal{M}_1$ and $\mathcal{M}_2$, we introduce the respective Kähler potentials K_1, K_2, Kähler metrics $g_{i\bar{j}}$, $g_{a\bar{b}}$, and symplectic invariant metrics $\mathbb{C}_1$, $\mathbb{C}_2$:

$$K_1 = -\log i \left(\overline{Z}^I \mathcal{G}_I - Z^I \overline{\mathcal{G}}_I \right), \qquad\qquad g_{i\bar{j}} = \partial_i \partial_{\bar{j}} K_1$$

$$K_2 = -\log i \left(\overline{X}^A \mathcal{F}_A - X^A \overline{\mathcal{F}}_A \right), \qquad\qquad g_{a\bar{b}} = \partial_a \partial_{\bar{b}} K_2$$

$$\mathbb{C}_{1\,\mathbb{IJ}} = \mathbb{C}_1^{\mathbb{IJ}} = \begin{pmatrix} 0 & \mathbb{1} \\ -\mathbb{1} & 0 \end{pmatrix} = \mathbb{C}_{2\,\mathbb{AB}} = \mathbb{C}_2^{\mathbb{AB}} \qquad \Rightarrow \qquad \mathbb{C}_{\mathbb{AB}} \mathbb{C}^{\mathbb{BC}} = -\delta_{\mathbb{A}}^{\mathbb{C}},$$

$$\mathbb{C}_{\mathbb{IJ}} \mathbb{C}^{\mathbb{JK}} = -\delta_{\mathbb{I}}^{\mathbb{K}}.$$

Finally, the coordinates $z^i \in \mathcal{M}_2$ are completed to coordinates on $\mathcal{M}_Q$ by the real axions $\xi^{\mathbb{I}} = (\xi^I, \tilde{\xi}_I)^T$ arising from the expansion of the higher dimensional RR potentials, together with the 4d dilaton φ and the axion a dual to the NSNS 2–form along the 4d spacetime.

The quaternionic manifold $\mathcal{M}_Q$ always admits a Heisenberg algebra of axionic isometries [22], which are gauged once fluxes are turned on in the higher-dimensional background [23, 24]. The NSNS fluxes, including geometric (and possibly non-geometric) fluxes, are encoded in a real 'bisymplectic' matrix $Q_{\mathbb{A}}^{\mathbb{I}}$, while the RR fluxes are encoded in a real symplectic vector $c^{\mathbb{A}}$. Explicitly,

$$Q_{\mathbb{A}}^{\mathbb{I}} = \begin{pmatrix} e_A{}^I & e_{AI} \\ m^{AI} & m^A{}_I \end{pmatrix}, \qquad\qquad c^{\mathbb{A}} = \begin{pmatrix} p^A \\ q_A \end{pmatrix}. \tag{2.77}$$

Abelianity and consistency of the gauging impose the following constraints on the charges [23, 24]

$$Q^T \mathbb{C}_2 Q = Q\,\mathbb{C}_1 Q^T = c^T Q = 0. \tag{2.78}$$

In the context of dimensional reductions, these can be traced back to the Bianchi identities for the higher-dimensional field strengths, together with the nilpotency of the exterior derivative on the compact manifold [26, 28] (cf. Appendix A.1).

The scalar potential generated by the gauging can be written as the sum of two contributions: $V = V_{\text{NS}} + V_{\text{R}}$, where V_{NS} can be seen to come from the reduction of the NSNS sector of type II theories, while V_{R} derives from the RR sector. Both these contributions take a symplectically invariant form, and read [24]

$$V_{\text{NS}} = -2e^{2\varphi}\Big[\overline{\varPi}_1^T \widetilde{Q}^T \mathbb{M}_2 \widetilde{Q} \varPi_1 + \overline{\varPi}_2^T Q \mathbb{M}_1 Q^T \varPi_2$$

$$+ 4\overline{\varPi}_1^T \mathbb{C}_1^T Q^T (\varPi_2 \overline{\varPi}_2^T + \overline{\varPi}_2 \varPi_2^T) Q \mathbb{C}_1 \varPi_1\Big]$$

$$V_{\text{R}} = -\tfrac{1}{2} e^{4\varphi} (c + \widetilde{Q}\,\xi)^T \mathbb{M}_2 (c + \widetilde{Q}\,\xi), \tag{2.79}$$

with

$$\widetilde{Q}_{\mathbb{I}}^{\mathbb{A}} = (\mathbb{C}_2^T Q\,\mathbb{C}_1)_{\mathbb{I}}^{\mathbb{A}}, \tag{2.80}$$

while $(\mathbb{M}_1)_{\mathbb{I}\mathbb{J}}$ and $(\mathbb{M}_2)_{\mathbb{A}\mathbb{B}}$ are symmetric, negative-definite matrices built respectively from the period matrices $(\mathcal{N}_1)_{IJ}$ and $(\mathcal{N}_2)_{AB}$ of the special Kähler geometries on $\mathcal{M}_1$ and $\mathcal{M}_2$ via the relation

$$\mathbb{M} = \begin{pmatrix} 1 & -\text{Re}\mathcal{N} \\ 0 & 1 \end{pmatrix} \begin{pmatrix} \text{Im}\mathcal{N} & 0 \\ 0 & (\text{Im}\mathcal{N})^{-1} \end{pmatrix} \begin{pmatrix} 1 & 0 \\ -\text{Re}\mathcal{N} & 1 \end{pmatrix}. \tag{2.81}$$

Nicely, expressions (2.79) for V_{NS} and V_{R} can be recast in a form that reminds that of the $\mathcal{N} = 2$ black hole potential, with the NSNS and RR fluxes $Q_{\mathbb{A}}^{\mathbb{I}}$ and $c^{\mathbb{A}}$ playing a role analogous to the black hole charges. The black hole central charge will here be replaced by the triplet of $\mathcal{N} = 2$ Killing prepotentials $\mathcal{P}_x$, $x = 1, 2, 3$ which describe the gauging under study. These read

$$\mathcal{P}_+ \equiv \mathcal{P}_1 + i\mathcal{P}_2 = 2e^{\varphi} \varPi_2^T Q\, \mathbb{C}_1 \varPi_1$$

$$\mathcal{P}_- \equiv \mathcal{P}_1 - i\mathcal{P}_2 = 2e^{\varphi} \varPi_2^T Q\, \mathbb{C}_1 \overline{\varPi}_1$$

$$\mathcal{P}_3 = e^{2\varphi} \varPi_2^T \mathbb{C}_2 (c + \widetilde{Q}\xi). \tag{2.82}$$

Here, $\mathcal{P}_\pm$ encode the contribution of the NSNS sector, while $\mathcal{P}_3$ describes the contribution of the RR sector [26, 28].

We find that the NSNS and RR potentials (2.79) can be recast in the suggestive form

$$V_{\text{NS}} = g^{a\bar{b}} D_a \mathcal{P}_+ \bar{D}_{\bar{b}} \overline{\mathcal{P}}_+ + g^{i\bar{j}} D_i \mathcal{P}_+ D_{\bar{j}} \overline{\mathcal{P}}_+ - 2|\mathcal{P}_+|^2$$

$$V_{\text{R}} = g^{a\bar{b}} D_a \mathcal{P}_3 \bar{D}_{\bar{b}} \overline{\mathcal{P}}_3 + |\mathcal{P}_3|^2, \tag{2.83}$$

whose main benefit is to involve only special Kähler covariant derivatives of the $\mathcal{P}_x$, which are defined as

$$D_i \mathcal{P}_x = (\partial_i + \tfrac{1}{2} \partial_i K_1) \mathcal{P}_x, \qquad D_a \mathcal{P}_x = (\partial_a + \tfrac{1}{2} \partial_a K_2) \mathcal{P}_x. \tag{2.84}$$

Equation (2.83) are the expressions for V_{NS} and V_{R} we are going to employ in the next sections. An expression closely related to the above rewriting of V_{NS} in terms of $\mathcal{P}_+$ appeared in [24], and our derivation of V_{R} in terms of $\mathcal{P}_3$ follows the same logic. In order to prove the equivalence between (2.79) and (2.83) we employ the following useful identities of special Kähler geometry [32]:

$$g^{i\bar{j}} D_i \Pi_1 D_{\bar{j}} \overline{\Pi}_1^T = -\tfrac{1}{2} \left(\mathbb{C}_1^T \mathbb{M}_1 \mathbb{C}_1 + i \mathbb{C}_1 \right) - \overline{\Pi}_1 \Pi_1^T \tag{2.85}$$

$$g^{a\bar{b}} D_a \Pi_2 D_{\bar{b}} \overline{\Pi}_2^T = -\tfrac{1}{2} \left(\mathbb{C}_2^T \mathbb{M}_2 \mathbb{C}_2 + i \mathbb{C}_2 \right) - \overline{\Pi}_2 \Pi_2^T. \tag{2.86}$$

These yield

$$g^{a\bar{b}} D_a \mathcal{P}_+ D_{\bar{b}} \overline{\mathcal{P}}_+ = -2 e^{2\varphi} \left(\overline{\Pi}_1^T \widetilde{Q}^T \mathbb{M}_2 \widetilde{Q} \Pi_1 + 2 \overline{\Pi}_1^T \mathbb{C}_1^T Q^T \Pi_2 \overline{\Pi}_2^T Q \mathbb{C}_1 \Pi_1 \right)$$

$$g^{i\bar{j}} D_i \mathcal{P}_+ D_{\bar{j}} \overline{\mathcal{P}}_+ = -2 e^{2\varphi} \left(\overline{\Pi}_2^T Q \mathbb{M}_1 Q^T \Pi_2 + 2 \overline{\Pi}_1^T \mathbb{C}_1^T Q^T \Pi_2 \overline{\Pi}_2^T Q \mathbb{C}_1 \Pi_1 \right)$$

$$-2|\mathcal{P}_+|^2 = -8 e^{2\varphi} \overline{\Pi}_1^T \mathbb{C}_1^T Q^T \overline{\Pi}_2 \Pi_2^T Q \mathbb{C}_1 \Pi_1$$

$$g^{a\bar{b}} D_a \mathcal{P}_3 D_{\bar{b}} \overline{\mathcal{P}}_3 + |\mathcal{P}_3|^2 = -\tfrac{1}{2} e^{4\varphi} (c + \widetilde{Q}\xi)^T \mathbb{M}_2 (c + \widetilde{Q}\xi), \tag{2.87}$$

and the equivalence between (2.79) and (2.83) is seen by addition of these four lines.

It is instructive to compare the quantities in (2.82) with the black hole central charge, which reads $Z_{\text{BH}} = \Pi_2^T \mathbb{C}_2 c$, where here $c = (p^A, q_A)$ is to be interpreted as the symplectic vector of electric and magnetic charges of the black hole. In addition to the fact that we are dealing with two quantities ($\mathcal{P}_+$ and $\mathcal{P}_3$) instead of a single one (Z_{BH}), we face here the further complication that these do not depend just on the covariantly holomorphic symplectic section Π_2 of the vector multiplet special Kähler manifold $\mathcal{M}_2$, but also on the scalars in the hypermultiplets. However, the constrained structure of the quaternionic manifold, determined via c-map from the special Kähler submanifold $\mathcal{M}_1$, comes to the rescue, yielding a relatively simple dependence of $\mathcal{P}_+$ and $\mathcal{P}_3$ on the 4d dilaton e^φ (appearing as a multiplicative factor) and on the axionic variables ξ^I (appearing in $\mathcal{P}_3$ as a scalar-dependent shift of the charge vector c). Finally, the $\mathcal{M}_1$ coordinates enter in $\mathcal{P}_+$ via the covariantly holomorphic symplectic section Π_1, making $\mathcal{P}_+$ a covariantly 'biholomorphic' object.

2.6.2 The Vacuum Equations

The extremization of the scalar potential (2.83) corresponds to the equations

$$\partial_\varphi V = 0 \iff V_{\text{NS}} + 2V_{\text{R}} = 0 \tag{2.88}$$

$$\partial_\xi V = 0 \iff \widetilde{Q}^T \mathbb{M}_2 \left(c + \widetilde{Q}\xi \right) = 0 \tag{2.89}$$

$$\partial_i V = 0 \iff iC_{ijk}g^{j\bar{j}}g^{k\bar{k}}\bar{D}_{\bar{j}}\mathcal{P}_-\bar{D}_{\bar{k}}\overline{\mathcal{P}}_+ - D_i\mathcal{P}_+\overline{\mathcal{P}}_+ + g^{a\bar{b}}D_aD_i\mathcal{P}_+\bar{D}_{\bar{b}}\overline{\mathcal{P}}_+ = 0 \tag{2.90}$$

$$\partial_a V = 0 \iff iC_{abc}g^{b\bar{b}}g^{c\bar{c}}\left(\bar{D}_{\bar{b}}\overline{\mathcal{P}}_-\bar{D}_{\bar{c}}\overline{\mathcal{P}}_+ + \bar{D}_{\bar{b}}\overline{\mathcal{P}}_3\bar{D}_{\bar{c}}\overline{\mathcal{P}}_3\right) - D_a\mathcal{P}_+\overline{\mathcal{P}}_+ + 2D_a\mathcal{P}_3\overline{\mathcal{P}}_3$$

$$+ g^{i\bar{j}}D_iD_a\mathcal{P}_+D_{\bar{j}}\overline{\mathcal{P}}_+ = 0. \tag{2.91}$$

To write (2.90), we used the following characterizing relations of special Kähler geometry [33]

$$D_iD_j\Pi_1 = iC_{ijk}g^{k\bar{k}}\bar{D}_{\bar{k}}\overline{\Pi}_1, \qquad D_i\bar{D}_{\bar{j}}\overline{\Pi}_1 = g_{i\bar{j}}\overline{\Pi}_1, \tag{2.92}$$

where C_{ijk} is the completely symmetric, covariantly holomorphic 3-tensor of the special Kähler geometry on $\mathcal{M}_1$. The analogous identities obtained by sending $1 \to 2$ and $i,j,k \to a,b,c$ have been used to derive Eq. (2.91). In particular, these relations imply

$$D_iD_j\mathcal{P}_+ = iC_{ijk}g^{k\bar{k}}\bar{D}_{\bar{k}}\mathcal{P}_-, \tag{2.93}$$

as well as

$$D_aD_b\mathcal{P}_+ = iC_{abc}g^{c\bar{c}}\bar{D}_{\bar{c}}\overline{\mathcal{P}}_- \qquad \text{and} \qquad D_aD_b\mathcal{P}_3 = iC_{abc}g^{c\bar{c}}\bar{D}_{\bar{c}}\overline{\mathcal{P}}_3. \tag{2.94}$$

The system of equations above, in particular Eqs. (2.90) and (2.91), take a form which reminds the attractor equations for black holes in $\mathcal{N} = 2$ supergravity. In the remaining of this section we will show how this set of equations is solved by a supersymmetry inspired set of first order conditions accounting for both supersymmetric and non-supersymmetric solutions.

2.6.3   First Order Conditions

In this subsection we propose a first order ansatz which generalizes the supersymmetry conditions and implies the vacuum equations (2.88)–(2.91). We refer the reader to [10] for the proof that vacuum equations are solved by this linear ansatz. We will later display some explicit examples of supersymmetric and non-supersymmetric vacua satisfying this first order ansatz.

We consider the following set of equations, linear in the $\mathcal{P}_x$ (and their derivatives), and hence in the flux charges:

$$\pm e^{\pm i\gamma - i\theta}\mathcal{P}_{\pm} = u\,\mathcal{P}_3$$

$$\pm e^{\pm i\gamma + i\theta}D_a\mathcal{P}_{\pm} = v\,D_a\mathcal{P}_3$$

$$D_i\mathcal{P}_+ \ = \ 0 \ = \ \bar{D}_i\mathcal{P}_-\,, \tag{2.95}$$

where γ, u, v, θ are real positive parameters. While γ will be just a free phase, the other three parameters will need to satisfy certain constraints given below. The ansatz (2.95) generalizes the AdS $\mathcal{N} = 1$ supersymmetry condition, which corresponds to the particular choice.[1]

$$\text{susy} \quad \Leftrightarrow \quad u = 2, \qquad v = 1, \qquad \theta = 0\,. \tag{2.96}$$

Our aim is to implement the first order ansatz (2.95) to derive non-supersymmetric solutions of the vacuum equations. For simplicity, we will restrict our analysis for the case in which $\mathcal{M}_2$ is a special Kähler manifold with a cubic prepotential. The ansatz (2.95) extremizes the scalar potential if the parameters u, v satisfy

$$\tfrac{1}{2}uv^2 - u^2v + u + v \ = \ 0\,, \tag{2.97}$$

and—under the assumption that $\mathcal{M}_2$ is cubic—if we further require that

$$D_a\mathcal{P}_3 \ = \ \alpha_3\,\partial_a K_2\,\mathcal{P}_3\,, \tag{2.98}$$

with α_3 given by

$$e^{3i\,\mathrm{Arg}(\alpha_3)+2i\,\mathrm{Arg}(\mathcal{P}_3)} = -\sqrt{\frac{4u}{3v}}\frac{1-v^2\,e^{2i\theta}}{2-uv\,e^{-2i\theta}}\,, \qquad |\alpha_3|^2 = \frac{u}{3v}\,. \tag{2.99}$$

---

[1] The $\mathcal{N} = 1$ conditions are completed by $\mathcal{P}_3 = -i\hat{\bar{\mu}}$, where $\hat{\mu} \neq 0$ is the parameter appearing in the Killing spinor equation on AdS, related to the AdS cosmological constant Λ via $\Lambda = -3|\hat{\mu}|^2$.

Notice that by evaluating the modulus square of both its sides, the first of (2.99) yields a constraint involving u, v, θ only:

$$4\,u\,v^2\cos(2\theta) \;=\; 3\,u^2v^3 - 4\,u\,v^4 - 4\,u + 12\,v\,. \qquad (2.100)$$

We remark that Eq. (2.95) can be rephrased in the symplectically covariant algebraic form

$$0 = Q^T \Pi_2 - i\,u\mathcal{P}_3 e^{i\theta-\varphi}\mathrm{Re}\!\left(e^{i\gamma}\Pi_1\right), \qquad (2.101)$$

$$0 = Q\mathbb{C}_1\mathrm{Re}\!\left(e^{i\gamma}\Pi_1\right), \qquad (2.102)$$

$$0 = 2Q\mathbb{C}_1\mathrm{Im}\!\left(e^{i\gamma}\Pi_1\right) + 2e^{-\varphi}(u+v)\mathrm{Re}\!\left[e^{-i\theta}\,\overline{\mathcal{P}}_3\mathbb{C}_2\Pi_2\right]$$
$$-e^{\varphi}(\mathbb{M}_2v\cos\theta - \mathbb{C}_2v\sin\theta)(c + \widetilde{Q}\xi). \qquad (2.103)$$

Notice that, precisely as in the supersymmetric case, the second equation actually follows from the first one.

2.6.3.1 Explicit Solutions

In the next section we will present explicit vacuum solutions **A**, **B**, **C** (cf. (2.125)–(2.127)) satisfying the linear ansatz, with the parameters u, v, θ given by

$$\mathbf{A}\ (\mathcal{N}=0): \quad u = 3v = 2\sqrt{\tfrac{6}{5}}, \qquad e^{i\theta} = \sqrt{\tfrac{1}{6}(\sqrt{5}-i)}$$

$$\mathbf{B}\ (\mathcal{N}=0): \quad u = v = 2, \qquad e^{i\theta} = \tfrac{1}{2}(1 - i\sqrt{3})$$

$$\mathbf{C}\ (\mathcal{N}=1): \quad u = 2v = 2, \qquad e^{i\theta} = 1\,.$$

2.6.4 Vacuum Solutions

In the following we present explicit $\mathcal{N}=0$ and $\mathcal{N}=1$ AdS$_4$ vacuum solutions of the $\mathcal{N}=2$ gauged supergravities under study.

For simplicity we focus on the case where the special Kähler scalar manifolds $\mathcal{M}_1$ and $\mathcal{M}_2$ are symmetric manifolds G/H with cubic prepotentials. The complete list of symmetric special Kähler manifolds with cubic prepotentials is given by the cosets G/H displayed in the first row of the following table (see e.g. [22])

$\frac{G}{H}$	$\frac{SU(1,1)}{U(1)}$	$\frac{SU(1,1)}{U(1)} \times \frac{SO(2,n)}{SO(n)\times U(1)}$	$\frac{Sp(6,\mathbb{R})}{SU(3)\times U(1)}$	$\frac{SU(3,3)}{SU(3)\times SU(3)\times U(1)}$	$\frac{SO^*(12)}{SU(6)\times U(1)}$	$\frac{E_{7(-25)}}{E_6\times SO(2)}$
$\mathcal{M}_Q$	$\frac{G_{2(2)}}{SO(4)}$	$\frac{SO(4,n+2)}{SO(n+2)\times SO(4)}$	$\frac{F_{4(4)}}{USp(6)\times SU(2)}$	$\frac{E_{6(2)}}{SU(6)\times SU(2)}$	$\frac{E_{7(-5)}}{SO(12)\times SU(2)}$	$\frac{E_{8(-24)}}{E_7\times SU(2)}$
$\mathbf{R}_G$	**2**	$(\mathbf{2, n+2})$	**14**	**20**	**32**	**56**
$\mathbf{R}_H$	**0**	$\mathbf{n+1}$	**6**	$\mathbf{(3,3)}$	**15**	**27**

The second row shows the quaternionic manifolds $\mathcal{M}_Q$ related to the special Kähler manifolds in the first row via the c-map [20]. The third row displays the symplectic G-representation under which the sections $\Pi_1^{\mathbb{I}}$ or $\Pi_2^{\mathbb{A}}$ transform. Finally, the last row shows the H-representation under which the scalar coordinates (z^i or x^a) on G/H transform.

We denote the cubic prepotentials on $\mathcal{M}_1$ and $\mathcal{M}_2$ respectively by

$$\mathcal{F} = \tfrac{1}{6}d_{abc}\frac{X^a X^b X^c}{X^0}, \qquad \mathcal{G} = \tfrac{1}{6}d_{ijk}\frac{Z^i Z^j Z^k}{Z^0}, \qquad (2.104)$$

where d_{abc} and d_{ijk} are scalar-independent, totally symmetric real tensors.

Choosing special coordinates, for $\mathcal{M}_2$ we have

$$X^A = (1, x^a), \qquad \mathcal{F}_A = (-f, f_a), \qquad f = \tfrac{1}{6}d_{abc}x^a x^b x^c,$$

$$f_a = \tfrac{1}{2}d_{abc}x^b x^c$$

$$\mathcal{V} = \tfrac{1}{6}d_{abc}x_2^a x_2^b x_2^c, \qquad \mathcal{V}_a = \tfrac{1}{2}d_{abc}x_2^b x_2^c, \qquad C_{abc} = e^{K_2}d_{abc}$$

$$K_2 = -\log(-8\mathcal{V}), \qquad \partial_a K_2 = \frac{i\mathcal{V}_a}{2\mathcal{V}}, \qquad (2.105)$$

where the complex coordinates x^a are split into real and imaginary parts as $x^a = x_1^a + i x_2^a$, with $x_2^a < 0$. Analogously, for $\mathcal{M}_1$ we have

$$Z^I = (1, z^i), \qquad \mathcal{G}_I = (-g, g_i), \qquad g = \tfrac{1}{6}d_{ijk}z^i z^j z^k,$$

$$g_i = \tfrac{1}{2}d_{ijk}z^j z^k$$

$$\tilde{\mathcal{V}} = \tfrac{1}{6}d_{ijk}z_2^i z_2^j z_2^k, \qquad \tilde{\mathcal{V}}_i = \tfrac{1}{2}d_{ijk}z_2^j z_2^k, \qquad C_{ijk} = e^{K_1}d_{ijk}$$

$$K_1 = -\log(-8\tilde{\mathcal{V}}), \qquad \partial_i K_1 = \frac{i\tilde{\mathcal{V}}_i}{2\tilde{\mathcal{V}}}. \qquad (2.106)$$

with $z_2^i < 0$. For the case of supergravity with no hypermultiplets other than the universal one, expressions (2.106) are replaced simply by $Z^0 = 1$, $\mathcal{G}_0 = -i$ and $e^{K_1} = \tfrac{1}{2}$.

Moreover, the following relations involving the metric $g_{a\bar{b}} = \partial_a \partial_{\bar{b}} K_2$ on $\mathcal{M}_2$ are valid:

$$g_{a\bar{b}} = \frac{1}{4\mathcal{V}^2}(\mathcal{V}_a \mathcal{V}_b - \mathcal{V}\mathcal{V}_{ab}), \qquad g^{a\bar{b}} = 2(x_2^a x_2^b - 2\mathcal{V}\mathcal{V}^{ab})$$

$$g^{ab}\mathcal{V}_b = 4\mathcal{V}x_2^a, \qquad g^{ab}\mathcal{V}_a \mathcal{V}_b = 12\mathcal{V}^2, \qquad (2.107)$$

where $\mathcal{V}_{ab} = d_{abc}x_2^c$, and $\mathcal{V}^{ab}$ is its inverse. Similar relations hold for the corresponding quantities on $\mathcal{M}_1$ with $a, b \to i, j$ and $\mathcal{V} \to \tilde{\mathcal{V}}$.

To perform the forthcoming computations, it is convenient to introduce the following holomorphic prepotentials $W_{\pm}$, W_3:

$$\mathcal{P}_{\pm} = e^{\frac{K_1 + K_2}{2} + \varphi} W_{\pm}, \qquad \mathcal{P}_3 = e^{\frac{K_2}{2} + 2\varphi} W_3, \qquad (2.108)$$

whose Kähler covariant derivatives are defined via

$$D_a W_x = (\partial_a + \partial_a K_2)W_x,$$

$$D_i W_+ = (\partial_i + \partial_i K_1)W_+, \qquad D_i W_- \equiv \partial_i W_- = 0,$$

$$D_{\bar{i}} W_- = (\partial_{\bar{i}} + \partial_{\bar{i}} K_1)W_-, \qquad D_{\bar{i}} W_+ \equiv \partial_{\bar{i}} W_+ = 0, \qquad (2.109)$$

with $\partial_a K_2$ and $\partial_i K_1$ given in (2.105) and (2.106) respectively. Explicitly, recalling (2.108) and (2.82), and taking $m^{AI} = m^A{}_I = 0$ for simplicity, one finds

$$W_+ = 2X^A(e_A{}^I \mathcal{G}_I - e_{AI}Z^I), \qquad W_- = 2X^A(e_A{}^I \bar{\mathcal{G}}_I - e_{AI}\bar{Z}^I),$$

$$W_3 = X^A(q_A + e_A{}^I \tilde{\xi}_I - e_{AI}\xi^I) - \mathcal{F}_A p^A, \qquad (2.110)$$

Recalling (2.105) and defining $f_{ab} = d_{abc}x^c$, we preliminarily compute:

$$\partial_a W_+ = 2(e_a{}^I \mathcal{G}_I - e_{aI}Z^I), \qquad \partial_a W_- = 2(e_a{}^I \bar{\mathcal{G}}_I - e_{aI}\bar{Z}^I),$$

$$\partial_a W_3 = q_a + e_a^I \tilde{\xi}_I - e_{aI}\xi^I + f_a p^0 - f_{ab}p^b. \qquad (2.111)$$

To solve the vacuum equations (2.88)–(2.91) in full generality is a challenging problem that goes beyond the scope of this paper. In the following we present some simple solutions as prototypes of the general case.

2.6.4.1 Only Universal Hypermultiplet

We start by considering the case of a gauged supergravity with a single hypermultiplet, which we identify with the universal hypermultiplet of string compactifications. Concerning the vector multiplets, we allow for an arbitrary number of them, and we just require that the associated special Kähler scalar manifold is symmetric and has a cubic prepotential,[2] specified by the 3-tensor d_{abc}. As it will be clear in the following, the latter assumption will allow us to perform computations in a more explicit fashion. In addition we assume that the only non-vanishing entries of the charge matrix Q be $e_{a0} \equiv e_a$. We remark that this choice might be generalized by using U-duality rotations. The constraints (2.78) require that $e_a p^a = 0$. For this choice of charges the prepotentials become

$$W_+ = W_- = -2e_a x^a ,$$

$$W_3 = q_0 + (q_a - e_a \xi)x^a + p^0 f - p^a f_a , \tag{2.112}$$

where all along this subsection we denote $\xi \equiv \xi^0$.

It is convenient to introduce the following shifted variables (assuming $p^0 \neq 0$):

$$\mathbf{x}^a = x^a - \frac{p^a}{p^0} , \qquad \mathbf{q}_0 = q_0 + \frac{q_a p^a}{p^0} - \frac{2P}{(p^0)^2} , \qquad \mathbf{q}_a = q_a - \frac{P_a}{p^0} , \tag{2.113}$$

with

$$P = \tfrac{1}{6} d_{abc} p^a p^b p^c , \qquad P_a = \tfrac{1}{2} d_{abc} p^b p^c . \tag{2.114}$$

In terms of these variables one finds

$$W_+ = W_- = -2e_a \mathbf{x}^a , \qquad\qquad \partial_a W_+ = \partial_a W_- = -2e_a ,$$

$$W_3 = \mathbf{q}_0 + (\mathbf{q}_a - \xi e_a)\mathbf{x}^a + p^0 \mathbf{f} , \qquad \partial_a W_3 = \mathbf{q}_a - \xi e_a + p^0 \mathbf{f}_a , \tag{2.115}$$

with

$$\mathbf{f} = \tfrac{1}{6} d_{abc} \mathbf{x}^a \mathbf{x}^b \mathbf{x}^c , \qquad \mathbf{f}_a = \tfrac{1}{2} d_{abc} \mathbf{x}^b \mathbf{x}^c . \tag{2.116}$$

In writing (2.115) we have used that $e_a p^a = 0$. Notice that since $\mathbf{x}_2^a \equiv x_2^a$, the expressions in (2.107) can be equivalently written with the bold variable. The advantage of using the bold variables introduced above is that the explicit dependence on p^a is entirely removed. Finally we introduce the following quantities built from the geometric fluxes e_a:

$$R = \tfrac{1}{6} d^{abc} e_a e_b e_c , \qquad R^a = \tfrac{1}{2} d^{abc} e_b e_c , \tag{2.117}$$

---

[2]In particular, the second property is relevant for type IIA compactifications on 6d manifolds M_6 with $SU(3)$ structure, where the Kähler potential K_2 is expected to take the cubic form $e^{-K_2} = \tfrac{4}{3} \int_{M_6} J \wedge J \wedge J$, where J is the almost symplectic 2–form on M_6. See Appendix A.1 for more details.

where d^{abc} is the contravariant tensor of the symmetric special Kähler geometry satisfying

$$d_{abc}\, d^{b(d_1 d_2}\, d^{d_3 d_4)c} = \tfrac{4}{3}\delta_a^{(d_1}\, d^{d_2 d_3 d_4)}. \tag{2.118}$$

Solutions of the vacuum equations can be found starting from the simple ansatz

$$\mathbf{x}^a = x\, R^a, \qquad \mathbf{q}_a = 0, \tag{2.119}$$

with $x = x_1 + ix_2$ a complex function of the charges to be determined. This ansatz can be motivated by noticing that once $\mathbf{q}_a$ is taken to zero, the only contravariant vector one can build with e_a's variables is R^a. Using (2.118) one finds the following relations:

$$d_{abc}\, R^b\, R^c = 2\, R\, e_a, \qquad R^a\, e_a = 3\, R,$$

$$\mathcal{V} = (x_2)^3 R^2, \qquad \mathcal{V}_a = (x_2)^2 R\, e_a, \qquad \mathbf{f} = x^3 R^2, \qquad \mathbf{f}_a = x^2 R\, e_a,$$

$$W_\pm = -6\, R\, x, \qquad W_3 = (\mathbf{q}_0 - 3R\,\xi\, x + p^0 R^2 x^3),$$

$$\partial_a W_+ = \partial_a W_- = -2 e_a, \qquad \partial_a W_3 = (\, p^0 R\, x^2 - \xi\,)e_a,$$

$$e^{-K_1} = 2, \qquad e^{-K_2} = -8\mathcal{V}. \tag{2.120}$$

Moreover, the covariant derivatives of the prepotentials take the form

$$D_a W_x = \frac{i\mathcal{V}_a}{2\mathcal{V}}\,\alpha_x W_x, \tag{2.121}$$

with

$$\alpha_\pm = 1 - \frac{2ix_2}{3x}, \qquad \alpha_3 = 1 - \frac{2ix_2 R(\,p^0 R\, x^2 - \xi\,)}{\mathbf{q}_0 - 3\, R\,\xi\, x + p^0 R^2 x^3}. \tag{2.122}$$

Using the relations (here there is no sum over x):

$$g^{a\bar{b}} D_a \mathcal{P}_x \bar{D}_{\bar{b}} \overline{\mathcal{P}}_x = 3|\alpha_x \mathcal{P}_x|^2,$$

$$iC_{abc}\, g^{b\bar{b}}\, g^{c\bar{c}}\, \bar{D}_{\bar{b}} \overline{\mathcal{P}}_x\, \bar{D}_{\bar{c}} \overline{\mathcal{P}}_x = 2\bar{\alpha}_x^2 \overline{\mathcal{P}}_x^2\, \frac{i\mathcal{V}_a}{2\mathcal{V}}, \tag{2.123}$$

the scalar potential reads

$$V = e^{K_1 + K_2 + 2\varphi}\,(3|\alpha_+|^2 - 2)|W_+|^2 + e^{K_2 + 4\varphi}\,(3|\alpha_3|^2 + 1)|W_3|^2. \tag{2.124}$$

Extermizing with respect to $x_{1,2}, \xi, \varphi$ one finds the solutions

- Solution **A** ($\mathcal{N} = 0$) :

$$x_1 = \xi = 0, \qquad x_2 = -5^{-1/6} \left(\frac{\mathbf{q}_0}{p^0 R^2} \right)^{1/3}, \qquad e^\varphi = \frac{1}{2\sqrt{2}} \, 5^{\frac{5}{6}} \left(\frac{R}{p^0 \mathbf{q}_0^2} \right)^{1/3},$$

$$V = -\frac{75}{64} \, 5^{\frac{5}{6}} \left(\frac{R^4}{p^0 \mathbf{q}_0^5} \right)^{1/3}. \tag{2.125}$$

- Solution **B** ($\mathcal{N} = 0$) :

$$x_1 = -\frac{1}{\sqrt{3}} x_2 = \left(\frac{\mathbf{q}_0}{20 p^0 R^2} \right)^{1/3}, \qquad \xi = \left(\frac{4 p^0 \mathbf{q}_0^2}{25 R} \right)^{1/3}, \qquad e^\varphi = \frac{\sqrt{2}}{\sqrt{3}} \left(\frac{25 R}{4 p^0 \mathbf{q}_0^2} \right)^{1/3},$$

$$V = -\frac{5}{\sqrt{3}} \left(\frac{25 R^4}{4 p^0 \mathbf{q}_0^5} \right)^{1/3}. \tag{2.126}$$

- Solution **C** ($\mathcal{N} = 1$) :

$$x_1 = \frac{1}{\sqrt{15}} x_2 = -\frac{1}{2} \left(\frac{\mathbf{q}_0}{20 p^0 R^2} \right)^{1/3}, \qquad \xi = -\left(\frac{p^0 \mathbf{q}_0^2}{50 R} \right)^{1/3}, \qquad e^\varphi = \frac{\sqrt{2}}{\sqrt{3}} \, 5^{\frac{1}{6}} \left(\frac{2R}{p^0 \mathbf{q}_0^2} \right)^{1/3},$$

$$V = -8\sqrt{3} \, 5^{-\frac{5}{6}} \left(\frac{2 R^4}{p^0 \mathbf{q}_0^5} \right)^{1/3}. \tag{2.127}$$

One also has to ensure $e^\varphi > 0$ and the positivity of the metric on the compact space M_6. The latter condition in this case amounts to $x_2^a < 0$. The above solutions generalize to an arbitrary number of vector multiplets the ones derived in [31] in the context of flux compactifications of massive type IIA on coset manifolds. Furthermore, they allow for non-vanishing charges p^a, q_a (satisfying $p^0 q_a = P_a$).

One can easily check that the solutions we found fall into the linear ansatz described in the last section with u, v, θ given by

$$\mathbf{A} \ (\mathcal{N} = 0): \quad u = 3v = 2\sqrt{\tfrac{6}{5}}, \qquad e^{i\theta} = \sqrt{\tfrac{1}{6}}(\sqrt{5} - i)$$

$$\mathbf{B} \ (\mathcal{N} = 0): \quad u = v = 2, \qquad e^{i\theta} = \tfrac{1}{2}(1 - i\sqrt{3})$$

$$\mathbf{C} \ (\mathcal{N} = 1): \quad u = 2v = 2, \qquad e^{i\theta} = 1.$$

2.6.4.2 Adding Hypermultiplets

The three solutions above can be generalized to the case of a cubic supergravity with arbitrary number of vector multiplets and hypermultiplets. For simplicity we focus again to the case where the vector multiplet scalar manifold is symmetric. Here we consider a vacuum configuration with non-trivial charges: $e_a{}^i$, $e_a \equiv e_{a0}$, p^0, $\mathbf{q}_0$, while $\mathbf{q}_a$ and all remaining charges in $Q_{\mathbb{A}}^{\mathbb{I}}$ are set to zero. Recalling (2.113), the prepotentials and their derivatives are now given by

$$W_+ = 2\mathbf{x}^a (e_a{}^i g_i - e_a), \qquad \partial_a W_+ = 2(e_a{}^i g_i - e_a), \qquad \partial_i W_+ = 2\mathbf{x}^a e_a{}^j g_{ij},$$

$$W_- = 2\mathbf{x}^a (e_a{}^i \bar{g}_i - e_a), \qquad \partial_a W_- = 2(e_a{}^i \bar{g}_i - e_a), \qquad \partial_{\bar{i}} W_- = 2\mathbf{x}^a e_a{}^j \bar{g}_{ij},$$

$$W_3 = \mathbf{q}_0 + \mathbf{x}^a (e_a{}^i \tilde{\xi}_i - e_a \xi^0) + p^0 \mathbf{f}, \qquad \partial_a W_3 = e_a{}^i \tilde{\xi}_i - e_a \xi^0 + p^0 \mathbf{f}_a,$$

$$(2.128)$$

where $g_{ij} = d_{ijk} z^k$. Again we follow an educated ansatz

$$\mathbf{x}^a = x\, R^a, \qquad z^i = z\, S^i, \qquad \tilde{\xi}_i = \zeta\, T_i, \qquad (2.129)$$

where we defined

$$R^a = \tfrac{1}{2} d^{abc} e_b e_c, \qquad R = \tfrac{1}{6} d^{abc} e_a e_b e_c$$

$$S^i = e_a{}^i R^a, \qquad T = \tfrac{1}{6} d_{ijk} S^i S^j S^k, \qquad T_i = \tfrac{1}{2} d_{ijk} S^j S^k. \qquad (2.130)$$

In addition we impose the following relation between the NSNS charges

$$d_{ijk} e_a{}^i e_b{}^j e_c{}^k = \beta\, d_{abc} \qquad \Rightarrow \qquad T_i e_a{}^i = \beta\, R\, e_a, \qquad T = \beta\, R^2, \qquad (2.131)$$

for a positive number β. With these assumptions, one has the following simplifications

$$\mathcal{V} = (x_2)^3 R^2, \qquad \mathcal{V}_a = (x_2)^2 R\, e_a, \qquad \mathbf{f} = x^3 R^2, \qquad \mathbf{f}_a = x^2 R\, e_a,$$

$$\tilde{\mathcal{V}} = (z_2)^3 T, \qquad \tilde{\mathcal{V}}_i = (z_2)^2 T_i, \qquad g = z^3 T, \qquad g_i = z^2 T_i$$

$$W_+ = 6\, R\, x\, (z^2 \beta R - 1), \qquad \partial_a W_+ = 2(z^2 \beta R - 1) e_a, \qquad \partial_i W_+ = 4\, x\, z\, T_i$$

$$W_3 = \mathbf{q}_0 - 3\, R\, x\, \hat{\xi} + p^0 R^2 x^3, \qquad \partial_a W_3 = (p^0 R\, x^2 - \hat{\xi}) e_a, \qquad (2.132)$$

where we introduced

$$\hat{\xi} = \xi^0 - \beta R\, \zeta. \qquad (2.133)$$

The new equations in the presence of additional hypermultiplets can be solved by taking

$$z = -i \sqrt{\frac{3}{\beta R}} \ . \tag{2.134}$$

Plugging this into $W_\pm$ and $D_a W_\pm$ one gets $W_\pm = -24xR$ and $D_a W_\pm = -8e_a$. Comparing with (2.120), one finds that the Killing prepotentials $\mathcal{P}_x$ are related to those found in the last section after the replacements

$$\varphi \rightarrow \hat{\varphi} \qquad \xi \rightarrow \hat{\xi} \qquad V \rightarrow e^{2c} V \tag{2.135}$$

with

$$e^{2\varphi} = e^{2\hat{\varphi}+c} \qquad \hat{\xi} = \xi^0 - \beta R \zeta \qquad e^c = 64 \, e^{K_1} = \left(\frac{64 \, \beta}{3^3 \, R}\right)^{\frac{1}{2}} \tag{2.136}$$

More precisely the Killing potentials in the cases of $h_1 > 0$ and $h_1 = 0$ are related via $\mathcal{P}_x^{h_1>0}(\varphi, \xi) = e^c \, \mathcal{P}_x^{h_1=0}(\hat{\varphi}, \hat{\xi})$. This implies that the three solutions (2.125)–(2.127) generalize to the case of arbitrary number of vector and hypermultiplets after the replacements (2.135) leading to

$$x \sim \left(\frac{\mathbf{q}_0}{p^0 R^2}\right)^{1/3}, \qquad \hat{\xi} \sim \left(\frac{p^0 \mathbf{q}_0^2}{R}\right)^{1/3} \qquad e^\varphi \sim \left(\frac{R^{\frac{1}{4}} \beta^{\frac{3}{4}}}{p^0 \mathbf{q}_0^2}\right)^{\frac{1}{3}},$$

$$V \sim \left(\frac{R \, \beta^3}{p^0 \mathbf{q}_0^5}\right)^{\frac{1}{3}} \ . \tag{2.137}$$

where we omit numerical coefficients depending on the particular solution **A**, **B**, **C**. Notice that the combination $\xi^0 + \beta R \zeta$ is a flat direction of the potential.

2.6.5 U-Invariant Cosmological Constant

In this section, we propose a U-duality invariant formula for the dependence on NSNS and RR fluxes of the scalar potential V at its critical points. This defines the cosmological constant $\Lambda = V|_{\partial V=0}$.

As considered in the treatment above, the setup is the following. The vectors' and hypers' scalar manifolds are given by $\mathcal{M}_2 = G/H$ and $\mathcal{M}_Q$. Here, G/H is a *symmetric* special Kähler manifold with cubic prepotential (d-special Kähler space, see e.g. [22]), with complex dimension h_2, coinciding with the number of (abelian) vector multiplets. On the other hand, $\mathcal{M}_Q$ is a *symmetric* quaternionic manifold, with quaternionic dimension $h_1 + 1$, corresponding to the number of hypermultiplets.

The manifold $\mathcal{M}_Q$ is the c-map [20] of the symmetric d-special Kähler space $\mathcal{M}_1 = \mathcal{G}/\mathcal{H} \subsetneq \mathcal{M}_Q$, with complex dimension h_1. Thus, the overall U-duality group is given by

$$U \equiv G \times \mathcal{G} \subsetneq Sp\,(2h_2 + 2, \mathbb{R}) \times Sp\,(2h_1 + 2, \mathbb{R})\,. \tag{2.138}$$

The Gaillard-Zumino [34] embedding of G and $\mathcal{G}$ is provided by the symplectic representations $\mathbf{R}_G$ and $\mathbf{R}_{\mathcal{G}}$, respectively spanned by the symplectic indices $\mathbb{A}$ and $\mathbb{I}$. The RR fluxes sit in the $(2h_2 + 2)$ vector representation $\mathbf{R}_G$ (as above, $a = 1, \dots, h_2$ and $i = 1, \dots, h_1$ throughout)

$$c^{\mathbb{A}} \equiv \begin{pmatrix} p^0 \\ p^a \\ q_0 \\ q_a \end{pmatrix}, \tag{2.139}$$

whereas the NSNS fluxes fit into the $(2h_2 + 2) \times (2h_1 + 2)$ bi-vector representation $\mathbf{R}_G \times \mathbf{R}_{\mathcal{G}}$

$$Q^{\mathbb{A}\mathbb{I}} = \mathbb{C}_2^{\mathbb{A}\mathbb{B}} Q_{\mathbb{B}}^{\mathbb{I}} = \begin{pmatrix} m^{AI} & m_I^A \\ -e_A^I & -e_{AI} \end{pmatrix}. \tag{2.140}$$

A priori, in presence of $c^{\mathbb{A}}$ and $Q^{\mathbb{A}\mathbb{I}}$, various $(G \times \mathcal{G})$-invariants, of different orders in RR and NSNS fluxes, can be constructed. Below we focus our analysis on invariants of total order 4 and 16 in fluxes, which respectively turn out to be relevant for the U-invariant characterization of Λ for the solutions **A**, **B**, **C** of Sects. 2.6.4.1 and 2.6.4.2.

2.6.5.1 Only Universal Hypermultiplet

Special Kähler symmetric spaces are characterized by a constant completely symmetric symplectic tensor $d_{\mathbb{A}_1 \mathbb{A}_2 \mathbb{A}_3 \mathbb{A}_4}$. This tensor defines a quartic G-invariant $\mathcal{I}_4\,(c^4)$ given by

$$\mathcal{I}_4\,(c^4) = d_{\mathbb{A}_1 \mathbb{A}_2 \mathbb{A}_3 \mathbb{A}_4} c^{\mathbb{A}_1} \dots c^{\mathbb{A}_4} \tag{2.141}$$

$$= -(p^0 q_0 + p^a q_a)^2 + \tfrac{2}{3} d_{abc}\, q_0 p^a p^b p^c - \tfrac{2}{3} d^{abc}\, p^0 q_a q_b q_c$$

$$+ d_{abc} d^{aef}\, p^b p^c q_e q_f\,.$$

A similar definition holds for the symplectic tensor $d_{\mathbb{I}_1 \mathbb{I}_2 \mathbb{I}_3 \mathbb{I}_4}$ of the symmetric coset $\mathcal{G}/\mathcal{H}$.

For the explicit solutions we have found, we take RR fluxes $c^{\mathbb{A}}$ with $p^0 \neq 0$ and

$$q_a \equiv \frac{1}{2} \frac{d_{abc} \, p^b \, p^c}{p^0}. \tag{2.142}$$

Plugging this into (2.141) and using the relation (2.118) (holding in *homogeneous symmetric d*-special Kähler geometries) one finds

$$\mathcal{I}_4 \left(c^4 \right) = -\left(p^0 \right)^2 \mathbf{q}_0^2 \qquad \text{with} \qquad \mathbf{q}_0 \equiv q_0 + \frac{1}{6} \frac{d_{abc} \, p^a \, p^b \, p^c}{(p^0)^2}. \tag{2.143}$$

Notice that the full dependence on p^a is encoded in the shift $q_0 \to \mathbf{q}_0$ and therefore we can, without loosing in generality, restrict ourselves to the simple choice $c^{\mathbb{A}} = (p^0, 0, \mathbf{q}_0, 0)$. The NSNS and RR fluxes are then given by

$$c^{\mathbb{A}} \equiv \begin{pmatrix} p^0 \\ 0 \\ \mathbf{q}_0 \\ 0 \end{pmatrix}, \qquad Q_0^{\mathbb{A}} \equiv \begin{pmatrix} 0 \\ 0 \\ 0 \\ -e_a \end{pmatrix}, \qquad Q^{\mathbb{A}0} = 0. \tag{2.144}$$

Besides $\mathcal{I}_4(c^4)$ one can build the following non-trivial quartic invariant

$$\mathcal{I}_4(c \, Q^3) = d_{\mathbb{A}_1 \mathbb{A}_2 \mathbb{A}_3 \mathbb{A}_4} c^{\mathbb{A}_1} \, Q_0^{\mathbb{A}_2} \, Q_0^{\mathbb{A}_3} \, Q_0^{\mathbb{A}_4} = -\tfrac{1}{6} p^0 d^{abc} e_a e_b e_c = -p^0 R. \tag{2.145}$$

Let us remark that $\mathcal{I}_4(c \, Q^3)$ is also invariant under the group $SO(2) = U(1)$ which, due to the absence of special Kähler scalars z^i in the hypersector, gets promoted to global symmetry. $\mathscr{G} = SO(2)$ is thus embedded into the symplectic group $Sp(2, \mathbb{R})$ via its irrepr. **2**, through which it acts on symplectic sections $(Z^0, \mathcal{G}_0)$. The $SO(2)$-invariance of $\mathcal{I}_4(c \, Q^3)$ is manifest, because the latter depends on the $SO(2)$-invariant $\mathcal{I}_2((Q_{\mathbb{A}})^2) = (Q_{\mathbb{A}0})^2 + (Q_{\mathbb{A}}^0)^2 = (Q_{\mathbb{A}0})^2$ (no sum over $\mathbb{A}$ is understood).

It is easy to see that solutions **A**, **B**, **C** given in Eqs. (2.125)–(2.127) depend only on the two combinations $\mathcal{I}_4(c^4)$ and $\mathcal{I}_4(c \, Q^3)$ given in (2.141) and (2.145) respectively. Upgrading these combinations to their U-duality invariant forms we can then write a manifestly $(G \times U(1))$-invariant formula for the AdS cosmological constant at the critical points

$$\Lambda = V \sim -\frac{\mathcal{I}_4^{4/3} \left(c \, Q^3 \right)}{\left| \mathcal{I}_4 \left(c^4 \right) \right|^{5/6}} \sim \frac{Q^4}{c^2}. \tag{2.146}$$

Notice that RR and NSNS fluxes play very different roles in their contribution to Λ. Indeed, the cosmological constant grows quartically on NSNS fluxes and fall off quadratically on RR charges. It would be nice to understand whether this is a general scaling feature of the gauged supergravities under study.

2.6.5.2 Many Hypermultiplets

Next let us consider the case with arbitrary number of hypermultiplets. From
(2.137), it follows that the solutions in this case depend only on the combinations
$\mathcal{I}_4 = (p^0 \mathbf{q}_0)^2$ and $\mathcal{I}_{16} \sim (p^0)^4 R\beta^3 \sim c^4 Q^{12}$. In order to write the solutions in a
U-duality invariant form we should then find an invariant $\mathcal{I}_{16}$ built out of 12 Q's and
4 c's that reduce to $(p^0)^4 R\beta^3$ on our choice of RR and NSNS fluxes. The following
$(G \times \mathscr{G})$-invariant quantity does the job

$$
\mathcal{I}_{16}\left(c^4 Q^{12}\right) \equiv d_{\mathbb{I}_1\mathbb{I}_2\mathbb{I}_3\mathbb{I}_4} d_{\mathbb{I}_5\mathbb{I}_6\mathbb{I}_7\mathbb{I}_8} d_{\mathbb{I}_9\mathbb{I}_{10}\mathbb{I}_{11}\mathbb{I}_{12}} d_{\mathbb{A}_1\mathbb{B}_1\mathbb{B}_2\mathbb{B}_3} d_{\mathbb{A}_2\mathbb{B}_5\mathbb{B}_6\mathbb{B}_7} d_{\mathbb{A}_3\mathbb{B}_9\mathbb{B}_{10}\mathbb{B}_{11}} \times
$$

$$
\times d_{\mathbb{A}_4\mathbb{B}_4\mathbb{B}_8\mathbb{B}_{12}} c^{\mathbb{A}_1} c^{\mathbb{A}_2} c^{\mathbb{A}_3} c^{\mathbb{A}_4} Q^{\mathbb{B}_1\mathbb{I}_1} \ldots Q^{\mathbb{B}_{12}\mathbb{I}_{12}}. \tag{2.147}
$$

The explicit expression of $\mathcal{I}_{16}\left(c^4 Q^{12}\right)$ is rather intricate. Nevertheless, this formula
undergoes a dramatic simplification when considering the configuration of NSNS
and RR fluxes, supporting the solutions found in Sect. 2.6.4. As before we encode
the full dependence on p^a in the shift $q_0 \to \mathbf{q}_0$ and therefore we restrict ourselves
to the charge vector choice $c^{\mathbb{A}} = (p^0, 0, \mathbf{q}_0, 0)$. More precisely, we take NSNS and
RR fluxes with all components of $Q^{\mathbb{A}\mathbb{I}}$, $c^{\mathbb{A}}$ zero except for

$$
Q_a{}^i = -e_a{}^i, \qquad Q_{a0} = -e_a, \qquad c^0 = p^0, \qquad c_0 = \mathbf{q}_0, \tag{2.148}
$$

where $e_a{}^i$ satisfy the constraint (2.131) for some $\beta \in \mathbb{R}_+$. A simple inspection to
(2.147), shows that contributions to $\mathcal{I}_{16}$ come only from the components $d_{ijk}^0 = -\frac{1}{6} d_{ijk}$ of $d_{\mathbb{I}_1..\mathbb{I}_4}$ and $d_0^{b_1 b_2 b_3} = -\frac{1}{6} d^{abc}$ of $d_{\mathbb{A}\mathbb{B}_1\mathbb{B}_2\mathbb{B}_3}$. Indeed using (2.118) one finds

$$
\mathcal{I}_{16}\left(c^4 Q^{12}\right) = \gamma \left(p^0\right)^4 R\beta^3, \tag{2.149}
$$

with[3]

$$
\gamma \equiv \frac{1}{6^6} (d^{abc} d_{abc} + h_2 + 3)^3. \tag{2.150}
$$

We conclude that the explicit vacuum solutions **A**, **B**, **C** obtained in Sect. 2.6.4
can be written in a manifestly U-duality invariant form in terms of the U-duality
invariants $\mathcal{I}_4(c^4)$ and $\mathcal{I}_{16}(c^4 Q^{12})$. In particular the value of the cosmological
constant Λ is given by the $(G \times \mathcal{G})$-invariant formula

$$
\Lambda = V \sim \frac{\mathcal{I}_{16}^{1/3}\left(c^4 Q^{12}\right)}{|\mathcal{I}_4\left(c^4\right)|^{5/6}} \sim \frac{Q^4}{c^2}. \tag{2.151}
$$

---

[3]Interestingly, the quantity $d^{abc} d_{abc}$, appearing in (2.150) is related to the Ricci scalar curvature $\mathcal{R}$
of the vector multiplets' scalar manifold G/H, whose general expression for a d-special Kähler
space reads $\mathcal{R} = -(h_2 + 1) h_2 + d^{abc} d_{abc}$, see [35–37].

Appendix

A.1 Flux/Gauging Dictionary for IIA on SU(3) Structure

Gauged $\mathcal{N} = 2$ supergravities with a scalar potential of the form studied in this paper can be derived by flux compactifications of type II theories on $SU(3)$ and $SU(3) \times SU(3)$ structure manifolds. While we refer to the literature (see e.g. [25,27, 29–31, 38–40]) for a detailed study of such general $\mathcal{N} = 2$ dimensional reductions and the related issues,[1] in this appendix we provide a practical dictionary between the 10d and the 4d quantities, with a focus on the scalar potential derived from $SU(3)$ structure compactifications of type IIA. In particular, we illustrate how the expressions one derives for V_{NS} and V_{R} are consistent with the scalar potential (2.79) studied in the main text.

A.1.1 SU(3) Structures and Their Curvature

An $SU(3)$ structure on a 6d manifold M_6 is defined by a real 2–form J and a complex, decomposable[2] 3–form Ω, satisfying the compatibility relation $J \wedge \Omega = 0$ as well as the non-degeneracy (and normalization) condition

$$\tfrac{i}{8}\Omega \wedge \bar{\Omega} = \tfrac{1}{6}J \wedge J \wedge J = vol_6 \neq 0 \quad \text{everywhere}. \tag{A.1}$$

Ω defines an almost complex structure I, with respect to which is of type $(3,0)$. In turn, I and J define a metric on M_6 via $g = JI$. The latter is required to be positive-definite, and vol_6 above denotes the associated volume form.

$SU(3)$ structures are classified by their torsion classes W_i, $i = 1, \ldots 5$, defined via [46]:

$$dJ = \tfrac{3}{2}\mathrm{Im}(\overline{W}_1\Omega) + W_4 \wedge J + W_3$$

$$d\Omega = W_1 \wedge J \wedge J + W_2 \wedge J + \overline{W}_5 \wedge \Omega, \tag{A.2}$$

where W_1 is a complex scalar, W_2 is a complex primitive $(1,1)$–form (primitive means $W_2 \wedge J \wedge J = 0$), W_3 is a real primitive $(1,2) + (2,1)$–form (primitive $\Leftrightarrow W_3 \wedge J = 0$), W_4 is a real 1–form, and W_5 is a complex $(1,0)$–form.

Reference [47] provides a formula for the Ricci scalar R_6 in terms of the torsion classes. We will restrict to $W_4 = W_5 = 0$, in which case the formula is

$$R_6 = \tfrac{1}{2}\big(15|W_1|^2 - W_2 \lrcorner \overline{W}_2 - W_3 \lrcorner W_3\big). \tag{A.3}$$

---

[1] In this context, see also [41–45] for studies of compactifications preserving $\mathcal{N} = 1$.

[2] A p–form is decomposable if locally it can be written as the wedging of p complex 1–forms.

This can equivalently be expressed as

$$R_6 \, vol_6 = -\tfrac{1}{2}\big[\, dJ \wedge *dJ + d\Omega \wedge *d\bar{\Omega} - (dJ \wedge \Omega) \wedge *(dJ \wedge \bar{\Omega})\,\big], \quad (A.4)$$

as it can be seen recalling (A.2) and computing

$$d\Omega \wedge *d\bar{\Omega} = 12|W_1|^2 vol_6 - J \wedge W_2 \wedge \overline{W}_2$$

$$= \big(12|W_1|^2 + W_2 \lrcorner \overline{W}_2\big) vol_6$$

$$dJ \wedge *dJ = \big(9|W_1|^2 + W_3 \lrcorner W_3\big) vol_6$$

$$(dJ \wedge \Omega) \wedge *(dJ \wedge \bar{\Omega}) = 36|W_1|^2 vol_6 . \quad (A.5)$$

A.1.2   The Scalar Potential from Dimensional Reduction

The 4d scalar potential receives contributions from both the NSNS and the RR sectors of type IIA supergravity. These are respectively given by

$$V_{\mathrm{NS}} = \frac{e^{2\varphi}}{2\mathcal{V}} \int_{M_6} \big(\tfrac{1}{2} H \wedge *H - R_6 * 1\big)$$

$$= \frac{e^{2\varphi}}{4\mathcal{V}} \int_{M_6} \Big[H \wedge *H + dJ \wedge *dJ + d\Omega \wedge *d\bar{\Omega}$$

$$-(dJ \wedge \Omega) \wedge *(dJ \wedge \bar{\Omega})\Big], \quad (A.6)$$

$$V_{\mathrm{R}} = \frac{e^{4\varphi}}{2} \int_{M_6} \big(F_0^2 * 1 + F_2 \wedge *F_2 + F_4 \wedge *F_4 + F_6 \wedge *F_6\big), \quad (A.7)$$

and the total potential reads $V = V_{\mathrm{NS}} + V_{\mathrm{R}}$. In (A.6), H is the internal NSNS field-strength, $\mathcal{V} = \int_{M_6} vol_6$, and φ is the 4d dilaton $e^{-2\varphi} = e^{-2\phi}\mathcal{V}$, where we are assuming that the 10d dilaton ϕ is constant along M_6. The k–forms F_k appearing in expression (A.7) are the internal RR field strengths, satisfying the Bianchi identity $dF_k - H \wedge F_{k-2} = 0$. The F_6 form can be seen as the Hodge-dual of the F_4 extending along spacetime, and the term $F_6 \wedge *F_6$ arises in a natural way if one considers type IIA supergravity in its democratic formulation [48].

A.1.2.1   Expansion Forms

In order to define the mode truncation, we postulate the existence of a basis of differential forms on the compact manifold in which to expand the higher dimensional fields. For a detailed analysis of the relations that these forms need to satisfy in order that the dimensional reduction go through, see in particular [27].

We take $\omega_0 = 1$ and $\tilde{\omega}^0 = \frac{vol_6}{\mathcal{V}}$, and we assume there exist a set of 2–forms ω_a satisfying

$$\omega_a \wedge *\omega_b \;=\; 4\,g_{ab}\,vol_6, \qquad\qquad \omega_a \wedge \omega_b \;=\; -d_{abc}\tilde{\omega}^c, \qquad\qquad (A.8)$$

where g_{ab} should be independent of the internal coordinates, d_{abc} should be a constant tensor, and the dual 4–forms $\tilde{\omega}^a$ are defined as

$$\tilde{\omega}^a \;=\; -\tfrac{1}{4\mathcal{V}}g^{ab} * \omega_b\,. \qquad\qquad (A.9)$$

From the above relations, we see that

$$\omega_a \wedge \tilde{\omega}^b \;=\; -\delta_a^b\,\tilde{\omega}^0, \qquad\qquad \omega_a \wedge \omega_b \wedge \omega_c \;=\; d_{abc}\tilde{\omega}^0\,. \qquad\qquad (A.10)$$

We also assume the existence of a set of 3–forms $\alpha_I,\,\beta^I$, satisfying

$$\alpha_I \wedge \beta^J \;=\; \delta_I^J\,\tilde{\omega}^0. \qquad\qquad (A.11)$$

Adopting the notation $\omega^{\mathbb{A}} = (\tilde{\omega}^A, \omega_A)^T = (\tilde{\omega}^0, \tilde{\omega}^a, \omega_0, \omega_a)^T$ and $\alpha^{\mathbb{I}} = (\beta^I, \alpha_I)^T$, we see that the symplectic metrics $\mathbb{C}$ appearing in the main text are here given by

$$\mathbb{C}_1^{\mathbb{IJ}} \;=\; -\int \alpha^{\mathbb{I}} \wedge \alpha^{\mathbb{J}}, \qquad \mathbb{C}_2^{\mathbb{AB}} \;=\; -\int \langle \omega^{\mathbb{A}}, \omega^{\mathbb{B}} \rangle\,, \qquad\qquad (A.12)$$

where the antisymmetric pairing $\langle\,,\,\rangle$ is defined on even forms ρ,σ as $\langle \rho,\sigma \rangle = [\lambda(\rho) \wedge \sigma]_6$, with $\lambda(\rho_k) = (-)^{\frac{k}{2}}\rho_k$, k being the degree of ρ, and $[\;]_6$ selecting the piece of degree 6.

The basis forms are used to expand Ω as

$$\Omega \;=\; Z^I\alpha_I - \mathcal{G}_I\beta^I \;=\; e^{-\frac{K_1}{2}}\,\Pi_1^{\mathbb{I}}\alpha_{\mathbb{I}}\,, \qquad\qquad (A.13)$$

and J together with the internal NS 2–form B as:

$$J \;=\; v^a\omega_a, \quad B \;=\; b^a\omega_a \qquad\Rightarrow\qquad e^{-B-iJ} \;=\; X^A\omega_A - \mathcal{F}_A\tilde{\omega}^A \;=\; e^{-\frac{K_2}{2}}\,\Pi_2^{\mathbb{A}}\omega_{\mathbb{A}}\,, \qquad (A.14)$$

where in the last equalities we define $\alpha_{\mathbb{I}} = \mathbb{C}_{\mathbb{IJ}}\alpha^{\mathbb{J}} = (\alpha_I, -\beta^I)^T$ and $\omega_{\mathbb{A}} = \mathbb{C}_{\mathbb{AB}}\omega^{\mathbb{B}} = (\omega_A, -\tilde{\omega}^A)^T$, and we adopt the symplectic notation defined in (2.76). Here, $(Z^I, \mathcal{G}_I)$ and $(X^A, \mathcal{F}_A)$ represent the holomorphic sections on the moduli spaces of Ω and $B + iJ$ expanded as above, which (under some conditions [26–28]) indeed exhibit a special Kähler structure, and correspond respectively to the manifolds $\mathscr{M}_1$ and $\mathscr{M}_2$ of the main text. Notice that here $X^A \equiv (X^0, X^a) \equiv (1, x^a) = (1, -b^a - iv^a)$, while $\mathcal{F}_A = \frac{\partial\mathcal{F}}{\partial X^A}$, where the cubic holomorphic function

$\mathcal{F} = \frac{1}{6} d_{abc} \frac{X^a X^b X^c}{X^0}$ is identified with the prepotential on $\mathcal{M}_2$. The Kähler potentials on $\mathcal{M}_1$ and $\mathcal{M}_2$ are recovered from $K_1 = -\log i \int \Omega \wedge \bar{\Omega}$ and $K_2 = -\log \frac{4}{3} \int J \wedge J \wedge J$, the latter yielding the metric g_{ab} appearing in (A.8). Notice that (A.1) implies $e^{-K_1} = e^{-K_2} = 8\mathcal{V}$.

The matrices $\mathbb{M}$ defined in (2.81) are given by

$$\mathbb{M}_{1,\mathbb{IJ}} = -\int \alpha_{\mathbb{I}} \wedge *\alpha_{\mathbb{J}}, \qquad \mathbb{M}_{2,\mathbb{AB}} = -\sum_k \int (e^B \omega_{\mathbb{A}})_k \wedge *(e^B \omega_{\mathbb{B}})_k,$$

(A.15)

and from the second relation one finds that the period matrix $\mathcal{N}_2$ on $\mathcal{M}_2$ reads

$$\text{Re}\mathcal{N}_{AB} = -\begin{pmatrix} \frac{1}{3} d_{abc} b^a b^b b^c & \frac{1}{2} d_{abc} b^b b^c \\ \frac{1}{2} d_{abc} b^b b^c & d_{abc} b^c \end{pmatrix}, \qquad \text{Im}\mathcal{N}_{AB} = -4\mathcal{V} \begin{pmatrix} \frac{1}{4} + g_{ab} b^a b^b & g_{ab} b^b \\ g_{ab} b^b & g_{ab} \end{pmatrix},$$

(A.16)

which is in agreement with the expression derived from $\mathcal{F}$ via the standard formula [49]

$$N_{AB} = \overline{\mathcal{F}}_{AB} + 2i \frac{\text{Im}\mathcal{F}_{AD} X^D \text{Im}\mathcal{F}_{BE} X^E}{X^C \text{Im}\mathcal{F}_{CE} X^E}, \qquad \mathcal{F}_{AB} \equiv \frac{\partial^2 \mathcal{F}}{\partial X^A \partial X^B}. \quad (A.17)$$

Finally, we also require the following differential conditions on the basis forms:

$$d\omega_a = e_a^{\mathbb{I}} \alpha_{\mathbb{I}}, \qquad d\alpha^{\mathbb{I}} = e_a^{\mathbb{I}} \tilde{\omega}^a, \qquad d\tilde{\omega}^a = 0, \qquad (A.18)$$

where the $e_a^{\mathbb{I}} = (e_a{}^I, e_{aI})$ are real constants, usually called 'geometric fluxes'. Defining the total internal NS 3-form as $H = H^{\text{fl}} + dB$, and expanding its flux part as

$$H^{\text{fl}} = -e_0{}^I \alpha_I + e_{0I} \beta^I \equiv -e_0^{\mathbb{I}} \alpha_{\mathbb{I}}, \qquad (A.19)$$

with constant $e_0^{\mathbb{I}}$, we can define $e_A^{\mathbb{I}} = (e_0^{\mathbb{I}}, e_a^{\mathbb{I}})^T$, and thus fill in half of the charge matrix Q introduced in (2.77):

$$Q_{\mathbb{A}}^{\mathbb{I}} = \begin{pmatrix} e_A^{\mathbb{I}} \\ 0 \end{pmatrix}. \qquad (A.20)$$

As noticed in [28], more general matrices, involving the $m_A^{\mathbb{I}}$ charges as well, can be obtained by considering non-geometric fluxes, or $SU(3) \times SU(3)$ structure compactifications. The nilpotency condition $d^2 = 0$ applied to (A.18), together with the Bianchi identity $dH = 0$, translates into the constraint

$$e_A^{\mathbb{I}} e_{B\mathbb{I}} = 0 \qquad \text{with} \quad e_{AI} = \mathbb{C}_{\mathbb{IJ}} e_A^{\mathbb{J}}, \qquad (A.21)$$

which, taking into account (A.20), is consistent with (2.78).

In the following, by using the above relations we recast in turn expressions (A.6) and (A.7) for V_{NS} and V_{R} in terms of 4d degrees of freedom, and show their consistency with (2.79).

A.1.2.2 Derivation of V_{NS}

Recalling the expansions of J, H and Ω defined above, using the assumed properties of the basis forms, and adopting the notation introduced in (2.76), one finds

$$\int dJ \wedge *dJ = -v^a v^b\, e_a^{\mathbb{I}}\, \mathbb{M}_{1,\mathbb{I}\mathbb{J}} e_b^{\mathbb{J}}, \qquad \int H \wedge *H = -b^A b^B\, e_A^{\mathbb{I}}\, \mathbb{M}_{1,\mathbb{I}\mathbb{J}} e_B^{\mathbb{J}},$$

$$\int d\Omega \wedge *d\bar\Omega = \tfrac{e^{-K_1}}{4\mathcal{V}}\, \Pi_1^{\mathbb{I}} e_{a\mathbb{I}} g^{ab} e_{b\mathbb{J}} \overline{\Pi}_1^{\mathbb{J}}, \qquad \int (dJ \wedge \Omega) \wedge *(dJ \wedge \bar\Omega)$$

$$= \tfrac{e^{-K_1}}{\mathcal{V}}\, \Pi_1^{\mathbb{I}} e_{a\mathbb{I}} v^a v^b e_{b\mathbb{J}} \overline{\Pi}_1^{\mathbb{J}},$$

where we define $b^A = (-1, b^a)$. Plugging this into (A.6), we get the NSNS contribution to V, expressed in a 4d language:

$$V_{\text{NS}} = -\frac{e^{2\varphi}}{4\mathcal{V}} \left[X^A e_A^{\mathbb{I}}\, \mathbb{M}_{1,\mathbb{I}\mathbb{J}} e_B^{\mathbb{J}}\, \overline{X}^B - \tfrac{e^{-K_1}}{4\mathcal{V}}\, \Pi_1^{\mathbb{I}} e_{a\mathbb{I}} (g^{ab} - 4v^a v^b) e_{b\mathbb{J}} \overline{\Pi}_1^{\mathbb{J}} \right]. \qquad \text{(A.22)}$$

Recalling (A.16), noticing that $\tfrac{1}{4\mathcal{V}}(g^{ab} - 4v^a v^b) = -(\operatorname{Im}\mathcal{N}_2)^{-1\,ab} - 4e^{K_2}(X^a \overline{X}^b + \overline{X}^a X^b)$, and recalling that $e^{-K_1} = e^{-K_2} = 8\mathcal{V}$, we conclude that (A.22) is consistent with (2.79).

A.1.2.3 Derivation of V_{R}

We consider the modified field-strengths $G_k \equiv \left[e^{-B} F \right]_k$, which satisfy the Bianchi identity $dG_k - H^{\text{fl}} \wedge G_{k-2} = 0$, and we define the expansions

$$G_0 = p^0, \qquad\qquad G_2 = p^a \omega_a, \qquad\qquad A_3 = \xi^{\mathbb{I}} \alpha_{\mathbb{I}}$$

$$G_4 = G_4^{\text{fl}} + dA_3 = (q_a - e_{a\mathbb{I}} \xi^{\mathbb{I}}) \tilde\omega^a, \qquad G_6 = G_6^{\text{fl}} - H^{\text{fl}} \wedge A_3 = (q_0 - e_{0\mathbb{I}} \xi^{\mathbb{I}}) \tilde\omega^0.$$

The Bianchi identities then amount just to the following constraint among the charges

$$p^A e_A^{\mathbb{I}} = 0, \qquad\qquad\qquad \text{(A.23)}$$

which, recalling (A.20), gives the last equality in (2.78). Then the integral in (A.7) reads

$$\sum_k \int F_k \wedge *F_k \;=\; \sum_k \int (e^B G)_k \wedge *(e^B G)_k \;=\; (c + \widetilde{Q}\xi)^T \mathbb{M}_2 (c + \widetilde{Q}\xi)\,, \quad \text{(A.24)}$$

where for the second equality we use (A.15), and here $(c + \widetilde{Q}\xi)^{\mathbb{A}} = (p^A,\, q_A - e_{A\mathbb{I}}\xi^{\mathbb{I}})^T$. The expression for V_R we obtain is therefore consistent with (2.79).

References

1. S. Ferrara, R. Kallosh, A. Strominger, N=2 extremal black holes. Phys. Rev. **D52**, 5412–5416 (1995). http://arxiv.org/abs/hep-th/9508072 (hep-th/9508072)
2. A. Strominger, Macroscopic entropy of $N = 2$ extremal black holes. Phys. Lett. **B383**, 39–43 (1996). http://arxiv.org/abs/hep-th/9602111 (hep-th/9602111)
3. S. Ferrara, R. Kallosh, Supersymmetry and attractors. Phys. Rev. **D54**, 1514–1524 (1996). http://arxiv.org/abs/hep-th/9602136 (hep-th/9602136)
4. S. Ferrara, R. Kallosh, Universality of supersymmetric attractors. Phys. Rev. **D54**, 1525–1534 (1996). http://arxiv.org/abs/hep-th/9603090 (hep-th/9603090)
5. R. Kallosh, New attractors. J. High Energy Phys. **12**, 022 (2005). http://arxiv.org/abs/hep-th/0510024 (hep-th/0510024)
6. P.K. Tripathy, S.P. Trivedi, Non-supersymmetric attractors in string theory. J. High Energy Phys. **03**, 022 (2006). http://arxiv.org/abs/hep-th/0511117 (hep-th/0511117)
7. A. Sen, Black hole entropy function and the attractor mechanism in higher derivative gravity. J. High Energy Phys. **09**, 038 (2005). http://arxiv.org/abs/hep-th/0506177 (hep-th/0506177)
8. J.F. Morales, H. Samtleben, Entropy function and attractors for AdS black holes. J. High Energy Phys. **10**, 074 (2006). http://arxiv.org/abs/hep-th/0608044 (hep-th/0608044)
9. S. Ferrara, A. Marrani, J.F. Morales, H. Samtleben, Intersecting attractors. Phys. Rev. **D79**, 065031 (2009). http://arxiv.org/abs/0812.0050 (arXiv:0812.0050)
10. D. Cassani, S. Ferrara, A. Marrani, J. F. Morales, H. Samtleben, A Special road to AdS vacua. J. High Energy Phys. **1002** (2010) 027. http://arxiv.org/abs/0911.2708 (arXiv:0911.2708)
11. S. Bellucci, S. Ferrara, R. Kallosh, A. Marrani, Extremal black hole and flux vacua attractors. Lect. Notes Phys. **755**, 115–191 (2008). http://arxiv.org/abs/0711.4547 (arXiv:0711.4547)
12. A. Sen, Black hole entropy function, attractors and precision counting of microstates. Gen. Relat. Gravity **40**, 2249–2431 (2008). http://arxiv.org/abs/0708.1270 (arXiv:0708.1270)
13. D.Z. Freedman, S.S. Gubser, K. Pilch, N.P. Warner, Renormalization group flows from holography supersymmetry and a c-theorem. Adv. Theor. Math. Phys. **3**, 363–417 (1999). http://arxiv.org/abs/hep-th/9904017 (hep-th/9904017)
14. K. Goldstein, R.P. Jena, G. Mandal, S.P. Trivedi, A c-function for non-supersymmetric attractors. J. High Energy Phys. **02**, 053 (2006). http://arxiv.org/abs/hep-th/0512138 (hep-th/0512138)
15. A. Sen, Quantum entropy function from AdS(2)/CFT(1) correspondence. Int. J. Mod. Phys. **A24**, 4225–4244 (2009). http://arxiv.org/abs/0809.3304 (arXiv:0809.3304)
16. J.D. Brown, M. Henneaux, Central charges in the canonical realization of asymptotic symmetries: an example from three-dimensional gravity. Commun. Math. Phys. **104**, 207–226 (1986)
17. J.A. Strathdee, Extended poincare supersymmetry. Int. J. Mod. Phys. **A2**, 273 (1987)
18. A. Van Proeyen, Tools for supersymmetry. http://arxiv.org/abs/hep-th/9910030 (hep-th/9910030)
19. P.K. Townsend, P-brane democracy, http://arxiv.org/abs/hep-th/9507048 (hep-th/9507048)

20. S. Cecotti, S. Ferrara, L. Girardello, Geometry of type II superstrings and the moduli of superconformal field theories. Int. J. Mod. Phys. **A4**, 2475 (1989)
21. S. Ferrara, S. Sabharwal, Quaternionic manifolds for type II superstring vacua of calabi-yau spaces. Nucl. Phys. **B332**, 317 (1990)
22. B. de Wit, F. Vanderseypen, A. Van Proeyen, Symmetry structure of special geometries. Nucl. Phys. **B400**, 463–524 (1993). http://arxiv.org/abs/hep-th/9210068 (hep-th/9210068)
23. R. D'Auria, S. Ferrara, M. Trigiante, S. Vaula, Gauging the heisenberg algebra of special quaternionic manifolds. Phys. Lett. **B610**, 147–151 (2005). http://arxiv.org/abs/hep-th/0410290 (hep-th/0410290)
24. R. D'Auria, S. Ferrara, M. Trigiante, On the supergravity formulation of mirror symmetry in generalized Calabi-Yau manifolds. Nucl. Phys. **B780**, 28–39 (2007). http://arxiv.org/abs/hep-th/0701247 (hep-th/0701247)
25. S. Gurrieri, J. Louis, A. Micu, D. Waldram, Mirror symmetry in generalized Calabi-Yau compactifications. Nucl. Phys. **B654**, 61–113 (2003). http://arxiv.org/abs/hep-th/0211102 (hep-th/0211102)
26. M. Grana, J. Louis, D. Waldram, Hitchin functionals in N = 2 supergravity. J. High Energy Phys. **01**, 008 (2006). http://arxiv.org/abs/hep-th/0505264 (hep-th/0505264)
27. A.-K. Kashani-Poor, R. Minasian, Towards reduction of type II theories on SU(3) structure manifolds. J. High Energy Phys. **03**, 109 (2007). http://arxiv.org/abs/hep-th/0611106 (hep-th/0611106)
28. M. Grana, J. Louis, D. Waldram, SU(3)×SU(3) compactification and mirror duals of magnetic fluxes. J. High Energy Phys. **04**, 101 (2007). http://arxiv.org/abs/hep-th/0612237 (hep-th/0612237)
29. D. Cassani, A. Bilal, Effective actions and N=1 vacuum conditions from SU(3)×SU(3) compactifications. J. High Energy Phys. **09**, 076 (2007). http://arxiv.org/abs/0707.3125 (arXiv:0707.3125)
30. D. Cassani, Reducing democratic type II supergravity on SU(3)×SU(3) structures. J. High Energy Phys. **06**, 027 (2008). http://arxiv.org/abs/0804.0595 (arXiv:0804.0595)
31. D. Cassani, A.-K. Kashani-Poor, Exploiting N=2 in consistent coset reductions of type IIA. Nucl. Phys. **B817**, 25–57 (2009). http://arxiv.org/abs/0901.4251 (arXiv:0901.4251)
32. L. Andrianopoli, R. D'Auria, S. Ferrara, M. Trigiante, Extremal black holes in supergravity. Lect. Notes Phys. **737**, 661–727 (2008). http://arxiv.org/abs/hep-th/0611345 (hep-th/0611345)
33. L. Andrianopoli et al., N = 2 supergravity and N = 2 super yang-mills theory on general scalar manifolds: symplectic covariance, gaugings and the momentum map. J. Geom. Phys. **23**, 111–189 (1997). http://arxiv.org/abs/hep-th/9605032 (hep-th/9605032)
34. M.K. Gaillard, B. Zumino, Duality rotations for interacting fields. Nucl. Phys. **B193**, 221 (1981)
35. E. Cremmer, A. Van Proeyen, Classification of Kahler manifolds in N=2 vector multiplet supergravity couplings. Class. Quantum Gravity **2**, 445 (1985)
36. S. Bellucci, A. Marrani, R. Roychowdhury, On quantum special Kaehler geometry, http://arxiv.org/abs/0910.4249 (arXiv:0910.4249
37. A. Ceresole, S. Ferrara, A. Marrani, 4d/5d Correspondence for the black hole potential and its critical points. Class. Quantum Gravity **24** (2007) 5651–5666. http://arxiv.org/abs/0707.0964 (arXiv:0707.0964)
38. A. Tomasiello, Topological mirror symmetry with fluxes. J. High Energy Phys. **06**, 067 (2005). http://arxiv.org/abs/hep-th/0502148 (hep-th/0502148)
39. T. House, E. Palti, Effective action of (massive) IIA on manifolds with SU(3) structure. Phys. Rev. **D72**, 026004 (2005). http://arxiv.org/abs/hep-th/0505177 (hep-th/0505177)
40. M. Grana, J. Louis, A. Sim, D. Waldram, E7(7) formulation of N=2 backgrounds. J. High Energy Phys. **07**, 104 (2009). http://arxiv.org/abs/0904.2333 (arXiv:0904.2333)
41. I. Benmachiche, T.W. Grimm, Generalized N = 1 orientifold compactifications and the Hitchin functionals. Nucl. Phys. **B748**, 200–252 (2006). http://arxiv.org/abs/hep-th/0602241 (hep-th/0602241)

42. P. Koerber, L. Martucci, From ten to four and back again: how to generalize the geometry. J. High Energy Phys. **08**, 059 (2007). http://arxiv.org/abs/0707.1038 (arXiv:0707.1038)
43. L. Martucci, On moduli and effective theory of N = 1 warped flux compactifications. J. High Energy Phys. **05**, 027 (2009). http://arxiv.org/abs/0902.4031 (arXiv:0902.4031)
44. J.-P. Derendinger, C. Kounnas, P.M. Petropoulos, F. Zwirner, Superpotentials in IIA compact- ifications with general fluxes. Nucl. Phys. **B715**, 211–233 (2005). http://arxiv.org/abs/hep-th/ 0411276 (hep-th/0411276)
45. G. Villadoro, F. Zwirner, N = 1 effective potential from dual type-IIA D6/O6 orientifolds with general fluxes. J. High Energy Phys. **06**, 047 (2005). http://arxiv.org/abs/hep-th/0503169 (hep-th/0503169)
46. S. Chiossi, S. Salamon, The intrinsic torsion of SU(3) and G_2 structures. http://arxiv.org/abs/ math/0202282 (math/0202282)
47. L. Bedulli, L. Vezzoni, The Ricci tensor of SU(3)-manifolds. J. Geom. Phys. **57**, 1125 (2007). http://arxiv.org/abs/math/0606786 (math/0606786)
48. E. Bergshoeff, R. Kallosh, T. Ortin, D. Roest, A. Van Proeyen, New formulations of D=10 supersymmetry and D8-O8 domain walls. Class. Quantum. Gravity **18**, 3359–3382 (2001). http://arxiv.org/abs/hep-th/0103233 (hep-th/0103233)
49. B. Craps, F. Roose, W. Troost, A. Van Proeyen, What is special Kaehler geometry? Nucl. Phys. **B503**, 565–613 (1997). http://arxiv.org/abs/hep-th/9703082 (hep-th/9703082)

Chapter 3
Extremality, Holography and Coarse Graining

Joan Simón

I discuss some of the concepts at the crossroads of gravitational thermodynamics, holography and quantum mechanics. First, the origin of gravitational thermodynamics due to coarse graining of quantum information is exemplified using the half-BPS sector of $\mathcal{N} = 4$ SYM and its LLM description in type IIB supergravity. The notion of black holes as effective geometries, its relation to the fuzzball programme and some of the puzzles raising for large black holes are discussed. Second, the semiclassical analysis giving rise to the extremal black hole/CFT conjecture. The latter is examined from the AdS_3/CFT_2 perspective.

3.1 Introduction

The equivalence principle linked gravitational physics with the geometry of space-time [78]. The extensive research on solutions to Einstein's equations, or generalisations thereof, and the study of their properties gave rise to many interesting facts and puzzles, especially when interpreted in the light of other branches of physics

- The connection between *black hole* physics and *thermodynamics* [26, 34, 91]
- The existence of curvature singularities [93, 94, 132] and observer dependent horizons
- The quantum nature of spacetime and its emergence in the classical limit.

General Relativity is viewed as an effective field theory. This follows, for example, from its lack of renormalizability or the existence of singularities. It suggests that a proper understanding of gravity requires the identification of the relevant

J. Simón (✉)
School of Mathematics, University of Edinburgh, Mayfield Road, King's Buildings, Edinburgh, EH9 3JZ, UK
e-mail: J.Simon@ed.ac.uk

S. Bellucci (ed.), *Supersymmetric Gravity and Black Holes*, Springer Proceedings in Physics 142, DOI 10.1007/978-3-642-31380-6_3,
© Springer-Verlag Berlin Heidelberg 2013

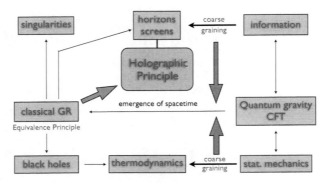

Fig. 3.1 Some of the concepts and relations discussed in these lectures

degrees of freedom in the ultraviolet (UV). The same conclusion may be reached using its connection to thermodynamics, through black hole physics. Thermodynamics is a *universal* branch of physics relatively independent of the microscopic details of the system under consideration. The birth of statistical mechanics, initiated with Boltzmann's work explaining the properties of macroscopic systems in thermal equilibrium in terms of the statistical averages of their microscopic degrees of freedom [44], further motivates the search for a quantum theory of gravity. The assignment of entropy to a classical spacetime raises the question as for what the microscopic degrees of freedom responsible for it are, i.e. what the analogue of the molecules in a gas is for spacetime.

The universality of gravity, in the sense that any energy source gravitates, may suggest that its proper formulation should follow from a first principle capturing the intricate structure one sees in its classical limit. This is what the *holographic principle* attempts [150, 152]. It states that the number of degrees of freedom N, understood as independent quantum states, describing a region B of spacetime, understood as an emergent structure from a more fundamental theory of matter and Lorentzian geometries, is bounded by the area $A(B)$ of its boundary ∂B

$$N \leq \frac{A(B)}{4G_N}, \tag{3.1}$$

where G_N stands for Newton's constant. For a review on the holographic principle, where a more mathematically accurate statement is given in terms of the *covariant entropy bound* [45] can be found in [46] (Fig. 3.1).

The holographic principle challenges the standard quantum field theory description of matter, stresses the non-local nature that gravity manifests in black hole physics, extends it to a general principle, going beyond the notion of event horizons, and emphasises that we can associate entropy and consequently, information, to *any* region of spacetime. Importantly, it does not provide an answer for what the degrees of freedom responsible for this entropy are.

String theory provides a mathematically well defined framework where to test some of these ideas. The importance of duality symmetries [99, 154], the discovery of D-branes as capturing non-perturbative aspects of string theory [134] and the formulation of the anti de Sitter (AdS) – conformal field theory (CFT) correspondence conjecture [85, 111, 155] are among the most important developments that have allowed to make both technical and conceptual progress in some of these issues. For example, string theory does provide with additional UV degrees of freedom, it allows to view certain black holes as bound states of D-branes [149] and the AdS/CFT correspondence provides an explicit realisation of the holographic principle itself.

In these lectures, I will mainly be concerned with

• The origin of gravitational thermodynamics in black hole physics through the coarse graining of quantum information and the use of the holographic principle to argue that such information loss is not necessarily confined to the black hole singularity, allowing us to view a black hole as a coarse grained object matching its standard thermal state interpretation. These ideas will be exemplified using the half-BPS sector of $\mathcal{N} = 4$ SYM and its LLM description in type IIB supergravity to describe the emergence of classical spacetime, singularities and entropy through coarse graining defined as a renormalization group (RG) transformation in a phase space description of quantum mechanics. The exposition will briefly cover the relation to the fuzzball programme, some speculative technical remarks on the information paradox [92] and will conclude with a discussion on some of the difficulties and puzzles appearing when trying to extend these ideas to large black holes.

• The description of the semiclassical methods that have recently been developed for *extremal* black holes in an attempt to understand more realistic black holes, explaining their macroscopic entropy, given by the universal Bekenstein-Hawking formula

$$S_{\text{BH}} = \frac{A}{4G_N},\qquad(3.2)$$

where A stands for the area of the black hole event horizon, in terms of the degrees of freedom living on a 2-dimensional (2d) CFT related to the black hole horizon, whose number of independent quantum states is universally controlled (at large temperatures) by Cardy's formula [52]

$$S_{\text{CFT}} = 2\pi\sqrt{\frac{c}{6}\left(L_0 - \frac{c}{24}\right)} + 2\pi\sqrt{\frac{\bar{c}}{6}\left(\bar{L}_0 - \frac{\bar{c}}{24}\right)},\qquad(3.3)$$

where c, $\bar{c}$ stand for the 2d non-chiral CFT central charges and $L_0 \pm \bar{L}_0$ are related to the conformal dimension and spin of the quantum states.

3.2   Black Holes, Thermodynamics and Fuzzball

The main goal of this section is to explore the *origin of gravitational thermodynamics* in the context of black hole physics, focusing in the relation between entropy and the emergence of spacetime and classical singularities through *coarse graining* of quantum information at microscopic scales.

The connection between black holes physics and thermodynamics has long been known [26, 34, 91]. The latter is a branch of physics dealing with systems of an effective infinite number of degrees freedom whose individual interactions are not measurable by a macroscopic observer. They are instead replaced by a coarse grained description involving an effective infinite reduction in the number of degrees of freedom at the price of introducing *entropy*, a magnitude measuring the amount of *information lost* in the reduction.

This last remark assumes the existence of a different physical description of the system at smaller scales not available to the macroscopic observer. One way to motivate quantum gravity is certainly to appeal to the universal link between statistical mechanics and thermodynamics when studying the black hole-thermodynamics relation. In black hole physics, it has long believed that the information loss about the true microscopic state of the system, responsible for the existence of entropy, is fully localised at the curvature singularity lying in the deep interior of the black hole. But this expectation is challenged by the holographic principle. Indeed, information takes space, and for a black hole, it involves a classical scale, the horizon scale. This would suggest that information about the state of the black hole, even if typically encoded in Planck scale (ℓ_p) physics, may be spread over macroscopic scales, such as the horizon scale, and not being merely localised to the singularity [121].

It is interesting to explore this observation a bit further. Whenever a quantum mechanical formulation is available, black holes are described by a *density matrix ρ*. The latter carries an intrinsic entropy

$$S_\rho = -\mathrm{Tr}\,(\rho \log \rho)\,. \tag{3.4}$$

This is a thermal description of the system, which differs from a microcanonical one, in which one would account for the black hole entropy by counting *pure states* $\{|\Psi_A\rangle\}$ having the same charges as the black hole. Quantum mechanically, density matrices and pure states are distinct. In principle, one can tell their difference apart by computing expectation values of operators $\mathcal{O}$

$$\mathrm{Tr}\,(\rho\,\mathcal{O}) \qquad \mathrm{vs} \qquad \langle\Psi_A|\mathcal{O}|\Psi_A\rangle\,. \tag{3.5}$$

But, how large are these differences? More importantly for the current discussion, do they manifest in classical gravitational physics? The answer to this question *must* necessarily be related to whether the information on the state of the black hole is fully encoded in the singularity or whether it spreads all the way to the horizon, as suggested by the holographic principle, though typically in ℓ_p cells.

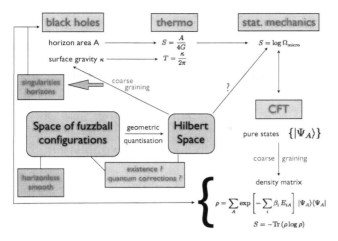

Fig. 3.2 From black holes to statistical mechanics and CFT and back using fuzzball ideas

The above digression suggests a very conservative approach to the origin of gravitational thermodynamics: the existence of some scale L such that after coarse graining all the microscopic information at smaller scales, all individual microstates would typically look alike. If true, a black hole should be understood as a *coarse grained* object, matching the standard density matrix description in quantum mechanics. Once more, one may be tempted to associate the scale L with ℓ_p, but this may well depend on the degeneracy of states encoded in the holographic relation dictated by the Bekenstein-Hawking formula (3.2). In string theory, if we generically denote the number of constituents of a given system by N, the scale that controls the classical gravitational curvature is proportional to $\ell_p N^{|\alpha|}$. This was checked by explicit calculations in some particularly symmetric configurations (see [121] and references therein) (Fig. 3.2).

Given the relation between black holes and thermal states, it is natural to wonder whether pure states allow any kind of reliable classical description in gravity. Generically, we would not expect this, but in the presence of enough supersymmetry, dynamics may be constrained enough so that as classical gravity becomes reliable, the information on some of these states may remain.[1] If these geometries exist, one would expect them to have the remarkable global property of being *horizonless*, to carry *no* entropy, matching their pure state nature.

This research programme was initiated in [108, 110]. The starting mathematical problem consists in determining the classical moduli space of configurations

---

[1]The language used in this argument may induce some readers to think of a transition between an open string (gauge theory) description to a closed string (gravitational) description. This may indeed be helpful, but the argument is more generic. If one assumes the existence of a fully quantum mechanical description of gravity, there is no guarantee that a typical pure state in such Hilbert space allows a reliable description in terms of a classical geometry when taking the classical limit.

having the same asymptotics and charges as a black hole, but having no horizon, being smooth and free of causal closed curves. Any configuration satisfying these requirements will be referred to as a *fuzzball* configuration. There exists an extensive literature on the subject. I refer the readers to some excellent reviews (and references therein): for the importance of fuzzball ideas to the resolution of the information paradox, see [120,124]; for the construction and interpretation of supergravity multi-center configurations, see [36, 146]; for the use of phase space quantisation, the importance of typicality and the nature of black holes as effective geometries, see [17, 146]. The intuitive idea consists in searching for configurations with the same asymptotics as a given black hole, but whose interior differs by removing the horizon and replacing the singularity by a smooth capped space. In some vague sense, the potentially complicated structure emerging in this inner region is reminiscent of Wheeler's ideas [11, 153].

Even if such classical moduli space exists, the connection to black hole entropy is still not apparent. This requires to quantise this moduli space, constructing a Hilbert space enabling us to count states with a given set of charges. This step can in principle be achieved through *geometric quantisation*, following Crnkovic and Witten [64] by quantising the phase space of such configurations. This is the approach reviewed in [17]. A priori, there is no guarantee this programme may work in a general non-extremal situation. On the contrary, the addition of generic non-extremal excitations may suggest the appearance of singularities in classical gravity. But, for highly supersymmetric configurations, where these solutions have been constructed explicitly, it has provided important insights into the nature of gravitational thermodynamics and the resolution of the information paradox.

In the following, we first discuss a universal statistical feature emerging when describing the differences between pure states quantum mechanically, correlating the difficulty in telling the individual states apart with the entropy of the system. Then, we will review the half-BPS sector of $\mathcal{N} = 4$ SYM and type IIB supergravity in $\mathrm{AdS}_5 \times \mathrm{S}^5$ where the amount of supersymmetry will allow us to explore the emergence of classical spacetime and singularities as quantum information about the precise state of the system is lost. This discussion will provide explicit evidence that in the case $L \to 0$ there exists an intricate "spacetime" fuzzball, which generically does *not* allow for a reliable classical description, but which can effectively be replaced by a singular configuration whose quantum mechanical description agrees with a coarse grained description of the exact quantum mechanical description of the system and which can not be told apart, by a semiclassical observer, from a typical pure state characterised by statistical mechanics considerations. We will conclude with a discussion regarding the important difficulties emerging when trying to extend these ideas to large black holes, i.e. $L \gg \ell_p$.

3.2.1 Distinction of States, Typicality and Fuzzballs

By first principles, quantum mechanical density matrices and pure states are different. Their differences can be encoded in expectation values (observables).

Assume a large black hole allows a quantum mechanical description. There will exist a large degeneracy of pure microstates encoded in its entropy S. Consequently, there must exist an statistical description of these states. The latter should allow us to mathematically determine the notion of *typical* state, i.e. how most of these degenerate pure microstates look like. Even more, in principle, it should be possible to determine the typical differences in observables among these typical pure states and the thermal state (density matrix), in an statistical sense.

These are quantum mechanical questions. Here one will also be interested in encoding the amount of information that survives the classical limit giving rise to a spacetime description that one calls black hole. This seems particularly relevant given the semiclassical nature of the calculations usually used to probe black hole physics. The main idea behind these statistical considerations is to provide some mathematically sounded ground where to compute the deviations from the averaged thermal answers and interpret them gravitationally: the black hole, viewed as a coarse grained effective geometry, will capture the averaged observable answer, but will differ from the exact quantum mechanical one in a given pure microstate. One would like to know by how much and what the most efficient observables are to highlight these differences.

It should not come as a surprise, that such deviations are highly suppressed. It was shown in [18] that the variances in local observables over the relevant Hilbert space of pure states are suppressed by a power of e^{-S}. To see this, consider a basis of the quantum mechanical Hilbert space of states with energy eigenvalues between E and $E + \Delta E$

$$\mathcal{M}_{\text{bas}} = \{ \, |s\rangle \; : \; H|s\rangle = e_s|s\rangle \; ; \; E \leq e_s \leq E + \Delta E \} \, . \tag{3.6}$$

The full set of states in this sector of the theory is

$$\mathcal{M}_{\text{sup}} = \left\{ |\Psi\rangle = \sum_s c_s^\psi |s\rangle \, , \quad \sum_s |c_s|^2 = 1 \right\} . \tag{3.7}$$

If the entropy of the system is $S(E)$, then the basis in (3.6) has dimension $e^{S(E)}$:

$$1 + \dim \mathcal{M}_{\text{sup}} = |\mathcal{M}_{\text{bas}}| = e^{S(E)}. \tag{3.8}$$

Take any local operator $\mathcal{O}$ and compute local observables of the form

$$c_\psi^k(x^1, \ldots, x^k) = \langle \Psi | \mathcal{O}(x^1) \cdots \mathcal{O}(x^k) | \Psi \rangle \, . \tag{3.9}$$

To measure how these vary over the ensemble $\mathcal{M}_{\text{sup}}$, define their averages and variances as

$$\langle c(x^1, \ldots, x^k)\rangle_{\mathcal{M}_{\text{sup}}} = \int D\Psi \, c_\psi(x^1, \ldots, x^k) \tag{3.10}$$

$$\text{var}[c(x^1, \ldots, x^k)]_{\mathcal{M}_{\text{sup}}} = \int D\Psi \left[(c_\psi(x^1, \ldots, x^k))^2 - \langle c(x^1, \ldots, x^k)\rangle^2_{\mathcal{M}_{\text{sup}}} \right]$$

The differences between states in $\mathcal{M}_{\text{sup}}$ in their responses to $\mathcal{O}$ will be captured by the standard-deviation to mean ratios

$$\frac{\sigma[c(x^1, \ldots, x^k)]_{\mathcal{M}_{\text{sup}}}}{\langle c(x^1, \ldots, x^k)\rangle_{\mathcal{M}_{\text{sup}}}} = \frac{\sqrt{\text{var}[c(x^1, \ldots, x^k)]_{\mathcal{M}_{\text{sup}}}}}{\langle c(x^1, \ldots, x^k)\rangle_{\mathcal{M}_{\text{sup}}}} \tag{3.11}$$

It was shown in [18] that

$$\text{var}[c(x^1, \ldots, x^k)]_{\mathcal{M}_{\text{sup}}} < \frac{1}{e^S + 1} \text{var}[c(x^1, \ldots, x^k)]_{\mathcal{M}_{\text{bas}}}. \tag{3.12}$$

This is a general result arising merely from statistical considerations. In particular, it is independent on the dynamical details of the theory. There are two ways to overcome this statistical suppression: given a fixed operator $\mathcal{O}$, one can wait for long time scales or given a fixed time scale, one can probe the state with operators having large statistical responses. Concerning the first option, it was pointed out that in lorentzian signature there was generically no time scale smaller than the Poincaré recurrence time to overcome these statistical factors [18], in agreement with previous claims in the literature [112]. On the other hand, extending previous work [104], it was shown that the analytic structure of these correlation functions, when working with euclidean signature in the complex plane, allowed the reduction of this time scale. It is unfortunately not clear whether these euclidean correlations can actually be measured by a single observer. The second option can provide a slightly different perspective on the comparison between *exact* quantum correlations and results derived from *semiclassical* considerations. In particular, the average correlation (3.10) will equal the thermal answer. The latter can be computed in a semiclassical approximation by doing a quantum field theory calculation in a black hole background. Our considerations above suggest

$$\langle \Psi | \Phi(x^1) \cdots \Phi(x^N) | \Psi \rangle_{\text{exact}} = \langle \Psi | \Phi(x^1) \cdots \Phi(x^N) | \Psi \rangle_{\text{eft}} + \mathcal{O}\left(e^{-(S-N)}\right) \tag{3.13}$$

This observation was also made in [3], building on [131]. It must be interpreted with care. It means that for a *fixed* time scale below the Poincaré recurrence, one should expect to find deviations from the thermal answer when the size of the probe operator is comparable to the entropy of the system. In the case of a CFT, size could stand for the conformal dimension of the probe operator, and operators generating black hole microstates themselves would be intuitive examples for this mechanism to be realised.

The importance of this statistical e^{-S} suppression in the context of the information paradox has been recently emphasised in [6]. Readers interested in a review on the information paradox itself and the perspective offered by the fuzzball ideas are referred to [122, 123]. For recent discussions regarding the physics of infalling observers in this context, see [125, 126].

These field theoretic considerations suggest that black hole backgrounds provide very accurate descriptions of the physics accessible to a classical observer [12, 13, 17, 109]. It is natural to ask, especially in an AdS/CFT context, whether the deviations in correlation functions mathematically described in terms of variances translate into some non-trivial spacetime scale and whether the latter survives the semiclassical limit. As mentioned before, a priori, this sounds improbable, since one would expect all these effects to be confined to the Planck scale. It is also technically hard to find reliable and precise results in the field theory side. Even if we restrict to highly supersymmetric sectors, the machinery for computing correlation functions in heavy states is still not fully developed, but there has been important progress towards achieving this goal in [62, 72, 73]. In the context of large black holes in the AdS$_5$/CFT$_4$ correspondence, these heavy states will have conformal dimension $\Delta \sim N^2$. Thus, even if working in a large N limit, the degeneracy of states is so large that standard large N perturbative diagrammatic counting arguments are not guaranteed to apply.

3.3 Half-BPS States in $\mathcal{N} = 4$ SYM Versus LLM Geometries

As an explicit example of the ideas outlined above, I will review the gauge and gravity descriptions for half-BPS states in $\mathcal{N} = 4$ Super Yang-Mills (SYM) and type IIB supergravity, respectively. These states are characterised by their R-charge J, since supersymmetry forces them to saturate the bound $\Delta = J$, where Δ is their conformal dimension. This set-up has two important advantages: its large amount of symmetry and its microscopic interpretation in terms of spherical rotating D-branes. The first guarantees that perturbative gauge theory states can be compared with their strongly coupled descriptions in supergravity. The second will help us to establish a dictionary between these two different descriptions. Unfortunately, the degeneracy of these states is not large enough to generate a macroscopic horizon. Hence, this sector of $\mathcal{N} = 4$ SYM will not be good enough to test our ideas for large black holes.

The particular half-BPS black holes we will be interested in were first found in [33]. These are type IIB $(U(1))^2 \times SO(4) \times SO(4)$ invariant supergravity configurations with constant dilaton and non-trivial metric and Ramond-Ramond (RR) four-form potential C_4

$$ds^2 = -\frac{\sqrt{\gamma}}{H} f \, dt^2 + \frac{\sqrt{\gamma}}{f} \, dr^2 + \sqrt{\gamma} \, r^2 \, ds_{S^3}^2 + \sqrt{\gamma} \, L^2 \, d\theta^2 + \frac{L^2}{\sqrt{\gamma}} \sin^2 \theta \, ds_{\tilde{S}^3}^2$$

$$+ \frac{1}{\sqrt{\gamma}} \cos^2 \theta \, [L \, d\phi + (H^{-1} - 1) \, dt]^2 \,, \tag{3.14}$$

$$C_4 = -\frac{r^4}{L} \gamma \, dt \wedge d^3 \Omega - L \, q \, \cos^2 \theta \, (L d\phi - dt) \,, \tag{3.15}$$

where $H = 1 + q/r^2$, $f = 1 + r^2 \, H/L^2$, $\gamma = 1 + q \sin^2 \theta / r^2$ and $L^4 = 4\pi \, g_s N \, l_s^4$ is the radius of AdS$_5$, with g_s the string coupling and l_s the string scale. These are asymptotically global AdS$_5 \times$S^5 singular configurations with vanishing horizon size carrying charge

$$\Delta = J = \omega \frac{N^2}{2} \,, \quad \omega = \frac{q}{L^2} \,, \tag{3.16}$$

These were coined *superstars* in [129], where they were interpreted as a distribution of giant gravitons [127]. A single giant graviton corresponds to a D3-brane wrapping $\tilde{S}^3$ while rotating at the speed of light in the ϕ direction. They preserve the same half of the supersymmetries as a point particle graviton, but they carry an R-charge of order N, i.e. $J \propto N$. Such N scaling is easy to understand: the dimensionless mass carried by the giant Δ_{giant} must be proportional to the D3-brane tension $T_{D3} = 1/(8\pi^3 g_s \, l_s^4)$ and its worldvolume

$$\Delta_{\text{giant}} \propto T_{D3} \, L^4 \propto N \,. \tag{3.17}$$

Physically, a pointlike graviton carrying R-charge of order N expands to an spherical D3-brane, through Myers' effect [128]. The solution (3.14)–(3.15) sources a certain number N_C of such giants that can be determined through the flux quantisation condition

$$N_C = \frac{1}{16\pi \, G_{10} T_{D3}} \int_{S^5} F_5 \, d^5 \Omega = \omega \, N \,, \tag{3.18}$$

where $F_5 = dC_4$ is the RR five-form field strength.

In the forthcoming sections, our main goal is to provide evidence that the superstar (3.14) corresponds to a coarse grained configuration emerging from integrating the quantum data of the exact quantum mechanical wave function.

3.3.1 Gauge Theory Description and Typicality

Due to the state–operator correspondence, highest weight half-BPS states in $\mathcal{N} = 4$ SYM correspond to multi-trace operators, $\mathcal{O} = \prod_{n,m} (\text{Tr} \, (X^m))^n$, built of a single real scalar field X transforming in the adjoint representation. In [41, 62], it was shown that the degrees of freedom describing these states are equivalent to N fermions $\{q_1, \ldots, q_N\}$ in a one dimensional harmonic potential. The number of these operators with conformal dimension $\Delta \sim N^2$ at very small chemical potential β [42, 102]

$$S_{1/2-\text{BPS}} \propto N \log N \,, \tag{3.19}$$

captures the large temperature behaviour of N harmonic oscillators plus an $1/N!$ statistical factor, in agreement with its fermionic interpretation. Given the exact nature of the half-BPS partition function, one can extrapolate this answer to strong coupling and estimate the size of an stretched horizon ρ_h by comparing the field theory entropy with the Bekenstein-Hawking entropy (3.2)

$$S_{\text{grav}} = S_{1/2-\text{BPS}} \sim N^2 \left(\frac{\rho_h}{L}\right)^3 \quad \Rightarrow \quad \frac{\rho_h}{L} \ll 1 \,. \tag{3.20}$$

One concludes the degeneracy of states is not large enough to generate a large horizon.

The fermionic description corresponds to an integrable system with ground state, the filled Fermi sea, consisting of fermions with energies $E_i^g = (i-1)\hbar + \hbar/2$ for $i = 1, \ldots, N$. Every excitation corresponds to a half-BPS state where the energy of the individual fermions is $E_i = f_i \hbar + \hbar/2$. Thus, there exists a correspondence between states and a set of unique non-negative ordered integers f_i. Exchanging these with a new set of integers $r_i = f_i - i + 1$, describing the excitations above the ground state, one establishes a correspondence between states and Young diagrams with N rows having as many boxes r_i as the excitation spectrum.

These diagrams are equally determined using the set of variables

$$c_N = r_1 \quad : \quad c_{N-i} = r_{i+1} - r_i \quad : \quad i = 1, 2, \ldots, (N-1) \,. \tag{3.21}$$

These c_j count the number of columns in the Young diagram of length j. These are particularly relevant when looking for a microscopic interpretation of gravity configurations with $\Delta \sim N^2$. This is because giant gravitons, as stressed before, have $\Delta \sim \mathcal{O}(N)$ and the operator dual to a single giant graviton is a subdeterminant [9]. In terms of Young diagrams, this operator corresponds to a single column. Thus, the number of columns in the diagram corresponds to the number of giant gravitons.

Given the large degeneracy of states having conformal dimension N^2, it is natural to use the statistical mechanics of the N fermions to identify how most of these states look like. Using the correspondence with Young diagrams, this will provide the shape of the typical diagram with these number of boxes. Since the superstar configuration (3.14) supports N_C giant gravitons, it is natural to consider an ensemble of Young diagrams in which the number of columns is held fixed [13].[2] Implementing this with a Lagrange multiplier λ, the canonical partition function equals

---

[2]There is more than one ensemble achieving this, see [71] for a discussion on this point.

$$Z = \sum_{c_1,c_2,\cdots,c_N=1}^{\infty} e^{-\beta \sum_j jc_j - \lambda(\sum_j c_j - N_C)} = \zeta^{-N_C} \prod_j \frac{1}{1 - \zeta q^j}, \quad \zeta = e^{-\lambda}.$$

(3.22)

The ensemble chemical potential β is fixed by requiring

$$\langle E \rangle = \Delta = q\partial_q \log Z(\zeta, q) = \sum_{j=1}^{N} \frac{j \zeta q^j}{1 - \zeta q^j},$$

(3.23)

whereas the Lagrange multiplier, or equivalently ζ, is fixed by

$$N_C = \sum_{j=1}^{N} \langle c_j \rangle = \sum_{j=1}^{N} \frac{\zeta q^j}{1 - \zeta q^j},$$

(3.24)

where we already computed the expected number of columns of length j, i.e. $\langle c_j \rangle$.

It is important to appreciate that the statistical properties of the system change considerably due to the constraint on the number of columns. Indeed, *without* the constraint, the amount of energy of the system, i.e. the number of boxes in the Young diagram, can grow without bound. Thus one expects the energy and entropy to increase monotonically as the system heats up. In the current set-up, this is no longer true. Our ensemble is characterised by the pair (N, N_C) and the conformal dimension Δ lies in a finite interval:

$$\langle E \rangle = \Delta \in [N_C, N \cdot N_C].$$

(3.25)

The lower bound corresponds to a single row Young diagram with N_C boxes; whereas the upper bound corresponds to a rectangular diagram in which all N rows have N_C boxes. Clearly, both bounds have a unique microstate, and so we can conclude our system has vanishing entropy in both situations.

Intuitively, one expects that as one increases the energy slightly above N_C, the entropy increases, heating up the system to a finite positive temperature. This behaviour should continue until the entropy reaches a maximum, which is achieved at $\Delta_s = N_C(N + 1)/2$, when half of the allowed boxes are filled in. If the energy goes beyond this value, we expect the entropy to start decreasing as the degeneracy of microstates will decrease. Indeed, there is a one-to-one map between Young diagrams with Δ boxes and $N_C(N + 1) - \Delta$ boxes. Thus, $S(\Delta) = S(N_C(N + 1) - \Delta)$. This shows that in this regime our system will have *negative* chemical potential. This discussion strongly suggests that the system achieves a vanishing chemical potential, i.e. $\beta \to 0$, when the entropy is maximised.

Let us examine different regimes in this ensemble, keeping N fixed and varying β. First, consider the *large-β* expansion. If the chemical potential is *positive*, this corresponds to the approximation $q = e^{-\beta} \ll 1$. In such regime, both sums in (3.23) and (3.24) are dominated by their first term. This allows us to conclude

$$\Delta \sim N_C \sim \langle c_1 \rangle = \frac{\zeta q}{1 - \zeta q} , \qquad (3.26)$$

which indeed corresponds to a diagram with a single row of N_C boxes since the dominant contribution comes from columns of length one.

If the chemical potential is *negative*, corresponding to the approximation $q = e^{-\beta} \gg 1$, the number of columns

$$N_C \sim \langle c_N \rangle = \frac{\zeta q^N}{1 - \zeta q^N} , \qquad (3.27)$$

is dominated by the number of columns of length N. This is so because the function $x/(1 - x)$ is monotonically increasing in x. Thus the corresponding diagram is a rectangular one, as expected. To further check this interpretation, we can compute the dominant contribution to the energy (3.23)

$$\begin{aligned}
\Delta &= \sum_{s=1}^{\infty} \zeta^s \sum_{j=1}^{N} j \, q^{sj} \\
&= \sum_{s=1}^{\infty} \zeta^s \left\{ q^s \frac{1 - q^{sN}}{1 - q^s} + q^{2s} \frac{1 - N \, q^{(N-1)s} + (N-1) q^{Ns}}{(1 - q^s)^2} \right\} \\
&\sim N \sum_{s=1}^{\infty} (\zeta \, q^N)^s = N \, N_C ,
\end{aligned} \qquad (3.28)$$

which indeed agrees with the energy of a rectangular diagram with N rows of N_C boxes each.

Consider now the small-β expansion. I shall treat both the positive and negative chemical potentials together including a sign in the definition of the expansion parameter (β). To prove the entropy goes through a maximum in this regime, we need to compute its expansion to second order in β. Using the identity,

$$S = \beta \Delta + \log Z = \beta \Delta - N_C \log \zeta - \sum_{j=1}^{N} \log(1 - \zeta \, q^j) , \qquad (3.29)$$

we realise we need to work out the expansions for ζ and Δ at second and first order, respectively. Expanding (3.24) and inverting the corresponding equation, we can find the expression for ζ:

$$\zeta(N, N_C) = \frac{\omega}{1 + \omega} \left(1 + \beta \frac{A_1}{N} + \beta^2 \frac{N+1}{12} ((N+2) - \omega(N-1)) \right) + \mathcal{O}(\beta^3) . \qquad (3.30)$$

In the above expression we have introduced the notation:

$$\omega \equiv \frac{N_C}{N}, \quad A_1 = \sum_{j=1}^{N} j = \frac{N(N+1)}{2}. \tag{3.31}$$

Expanding (3.23), the conformal dimension at first order is

$$\Delta = \omega A_1 - \beta \frac{A_1}{6} \omega (1 + \omega)(N - 1) + \mathcal{O}(\beta^2). \tag{3.32}$$

Notice that at large-N, the dominant energy is the one computed for the superstar in [129], suggesting that the $\beta \to 0$ limit the large-N typical state is described in spacetime as the superstar. We will explicitly show this later. The typical states in this vanishing chemical potential ensemble correspond to Young diagrams that are nearly triangular. That is, on average, there is a constant gap (ω) of energy between the excitations of the $(j + 1)$-st and j-th fermions, the first fermion having an average energy ω itself. Moreover, having computed the linear β dependence allows us to confirm that for positive β, the conformal dimension is smaller than the superstar energy, whereas for negative β, it can exceed the latter, in agreement with our general entropic arguments.

Finally, if the superstar ensemble maximises the entropy, the entropy expansion in β should have no linear dependence in it and its second order coefficient must be negative for all values of N, N_C. Carrying out the computation, we obtain

$$S = -\log \left(\frac{\omega^{N_C}}{(1 + \omega)^{N+N_C}} \right) - \beta^2 \frac{\omega(1 + \omega)}{24} N(N^2 - 1) + \mathcal{O}(\beta^4). \tag{3.33}$$

Since $N \geq 1$, the coefficient of β^2 is negative, for any value of (N, N_C), as expected. According to our discussion, we should identify the first term above as a microscopic derivation for the entropy of the superstar.

3.3.1.1 Limit Curve

Let us introduce two coordinates x and y along the rows and columns of the Young diagram. In my conventions, the origin $(0, 0)$ is the bottom left corner of the diagram, x increases going up while y increases to the right. In the fermion language, x labels the particle number and y its excitation above the vacuum. Determining the typical state consists in computing the curve $y(x)$ describing the shape of the Young diagram. This can be done by identifying this mathematical object with the expectation value

$$y(x) = \sum_{i=N-x}^{N} \langle c_i \rangle. \tag{3.34}$$

In the limit $\hbar \to 0$, $N \to \infty$, keeping the Fermi level $\hbar N$ fixed, we can treat x and y as continuum variables, replace the summation by an integral, and derive the limit curve [13]

$$y(x) = \frac{\log(1 - e^{-\beta N})}{\beta} - \frac{\log(1 - e^{-\beta(N-x)})}{\beta} . \qquad (3.35)$$

For a discussion concerning the size of the fluctuations, see [13, 71]. We will be particularly interested in the $\beta \to 0$ limit of this limit curve. Before studying this further, let us review the classical gravitational description for these half-BPS configurations.

3.3.2 Gravity Description

All the relevant $U(1) \times SO(4) \times SO(4)$ half-BPS supergravity backgrounds for our discussion were constructed in [105]. These involve a metric

$$ds^2 = -h^{-2}(dt + V_i dx^i)^2 + h^2(d\eta^2 + dx^i dx^i) + \eta\, e^G d\Omega_3^2 + \eta\, e^{-G} d\tilde{\Omega}_3^2 , \quad (3.36)$$

and a self-dual five-form field strength $F_{(5)} = F \wedge d\Omega_3 + \tilde{F} \wedge d\tilde{\Omega}_3$, where $d\Omega_3$ and $d\tilde{\Omega}_3$ are the volume forms of the two three-spheres where the two $SO(4)$s are geometrically realised. The full configuration is uniquely determined in terms of a *single* scalar function $z = z(\eta, x_1, x_2)$ satisfying the linear differential equation (see [105] for a complete discussion)

$$\partial_i \partial_i z + \eta \partial_\eta \left(\frac{\partial_\eta z}{\eta} \right) = 0 . \qquad (3.37)$$

Notice $\Phi(\eta; x^1, x^2) = z\eta^{-2}$ satisfies the Laplace equation for an electrostatic potential in six dimensions being spherically symmetric in four of them. The coordinates x_1, x_2 parametrize an $\mathbb{R}^2$, while η is the radial coordinate in the transverse $\mathbb{R}^4$ in this auxiliary six-dimensional manifold. Thus, the general solution

$$z(\eta; x_1, x_2) = \frac{\eta^2}{\pi} \int dx'_1\, dx'_2\, \frac{z(0; x'_1, x'_2)}{[(x - x')^2 + \eta^2]^2} , \qquad (3.38)$$

depends on the boundary condition $z(0; x_1, x_2)$. To be complete, we remind the reader of the relations determining the full configuration in terms of this scalar function

$$h^{-2} = 2\eta \cosh G, \qquad (3.39)$$

$$\eta \partial_\eta V_i = \epsilon_{ij} \partial_j z, \qquad \eta(\partial_i V_j - \partial_j V_i) = \epsilon_{ij} \partial_\eta z \qquad (3.40)$$

$$z = \frac{1}{2} \tanh G \tag{3.41}$$

$$F = dB_t \wedge (dt + V) + B_t dV + d\hat{B},$$

$$\tilde{F} = d\tilde{B}_t \wedge (dt + V) + \tilde{B}_t dV + d\hat{\tilde{B}} \tag{3.42}$$

$$B_t = -\frac{1}{4}\eta^2 e^{2G}, \qquad\qquad \tilde{B}_t = -\frac{1}{4}\eta^2 e^{-2G} \tag{3.43}$$

$$d\hat{B} = -\frac{1}{4}\eta^3 *_3 d(\frac{z+\frac{1}{2}}{\eta^2}), \qquad d\hat{\tilde{B}} = -\frac{1}{4}\eta^3 *_3 d(\frac{z-\frac{1}{2}}{\eta^2}) \tag{3.44}$$

It was shown in [105] that *regularity* forces the boundary condition

$$z(0; x_1, x_2) = \pm 1/2. \tag{3.45}$$

Introducing $u(0; x_1, x_2) = 1/2 - z(0; x_1, x_2)$, the energy and flux quantisation condition are

$$\Delta = \int_{\mathbb{R}^2} \frac{d^2x}{2\pi\hbar} \frac{1}{2} \frac{x_1^2 + x_2^2}{\hbar} u(0; x_1, x_2) - \frac{1}{2} \left(\int_{\mathbb{R}^2} \frac{d^2x}{2\pi\hbar} u(0; x_1, x_2) \right)^2, \tag{3.46}$$

$$N = \int_{\mathbb{R}^2} \frac{d^2x}{2\pi\hbar} u(0; x_1, x_2), \tag{3.47}$$

where $\hbar = 2\pi\ell_p^4$ due to the non-standard units carried by $\{x_1, x_2\}$. These expressions resemble *expectation values* computed in the phase space of one of the fermions appearing in our gauge theory discussion, suggesting the function $u(0; x_1, x_2)$ should be identified with the semiclassical limit of the quantum single-particle phase space distributions of the dual fermions.

3.3.3 Gauge-Gravity Correspondence and Coarse Graining

The classical moduli space of configurations described in [105] was geometrically quantised in [84, 118].[3] The Hilbert space they constructed when restricting to the subspace of BPS configurations was isomorphic to the one describing N fermions in a one dimensional harmonic oscillator appearing in the gauge theory. This automatically guarantees that the gauge theory and gravity counting of states match. In the following, we will focus on the information loss by coarse graining of quantum information.

---

[3] The same methods were applied to the D1–D5 system in [135].

The matching of the gauge theory and gravity descriptions requires a dictionary. Focusing on $U(1)$ invariant configurations in the x_1, x_2 plane, it was proposed in [13] that in the semiclassical limit $\hbar \to 0$ with $\hbar N$ fixed, the integral formulae (3.46) and (3.47) extend to differential relations

$$\frac{u(0; r^2)}{2\hbar} dr^2 = dx , \qquad \frac{r^2 u(0; r^2)}{4\hbar^2} dr^2 = (y(x) + x) dx . \qquad (3.48)$$

The first equation relates the number of particles in phase space within a band between r and $r + dr$ to the number of particles as determined by the rows of the associated Young diagram. The second equation matches the energy of the particles in phase space within a ring of width dr to the energy in terms of the Young diagram coordinates. This is equivalent to identifying the $y = 0$ plane in the bulk with the semiclassical limit of the phase space of a single fermion [84, 117, 151]. Combining both equations in (3.48), we find $y(x) + x = r^2/(2\hbar)$ and taking derivatives,

$$u(0; r^2) = \frac{1}{1 + y'} \qquad \Leftrightarrow \qquad z(0; r^2) = \frac{1}{2} \frac{y' - 1}{y' + 1} . \qquad (3.49)$$

This establishes a dictionary between the boundary condition $u(0; x_1, x_2)$ in supergravity and the slope $(y'(x) = dy/dx)$ of the Young diagram of the corresponding field theory state.

We will now rederive Eq. (3.49) from a different perspective, emphasising the intrinsic loss of information involved in the semiclassical limit we are taking. Given the exact quantum mechanical phase space distribution, one can seek a new distribution function that is sufficient to describe the effective response of coarse grained semiclassical observables in states that have a limit as $\hbar \to 0$. The latter will be called the *coarse grained* or *grayscale* distribution. Since we are interested in single Young diagram states, it can only depend on the energy $E = (p^2 + q^2)/2$. To derive this, let ΔE be a coarse graining scale such that $\Delta E/\hbar \to \infty$ in the $\hbar \to 0$ limit. The *grayscale* distribution $g(E)$ must equal the quotient of the number of fermions with energies between E and $E + \Delta E$[4]

$$R(E, \Delta E) = 2\pi\hbar \int_E^{E+\Delta E} dq\, dp\, \mathrm{Hu}(q, p), \qquad (3.50)$$

by the area of phase space between these scales

$$\mathrm{Area} = \int_E^{E+\Delta E} dq\, dp = 2\pi \Delta E . \qquad (3.51)$$

---

[4]We are using the Husimi distribution $\mathrm{Hu}(q, p)$ given its nice properties in the classical limit. For further discussion, see [13]. For a review on phase space distributions, see [98].

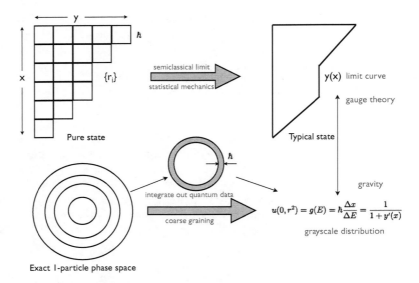

Fig. 3.3 From pure states, to typical states and their coarse grained phase space density description providing a gravity boundary condition

Hence, the *grayscale* distribution equals

$$g(E) = 2\pi\hbar \left[\frac{R(E, \Delta E)}{2\pi \Delta E} \right]. \tag{3.52}$$

Since the number of fermions in this area is related to the continuous coordinate x and the area to the energy, we conclude

$$g(E) = \hbar \frac{\Delta x}{\Delta E} = \frac{\hbar}{\partial E/\partial x} = \frac{1}{1 + y'}. \tag{3.53}$$

where the gauge-gravity proposal $E = \hbar(x + y(x))$ was used (Fig. 3.3).

Remarkably, this is precisely the quantity (3.49) that we proposed on general grounds to determine the classical supergravity half-BPS solution associated to a Young diagram. Thus, explicitly, the proposal is

$$u(0; r^2) = g(r^2/2) = \frac{1}{1 + y'} \tag{3.54}$$

for all half-BPS states that are described by single Young diagrams in the semiclassical limit.

3.3.3.1   Matching the Superstar Geometry

As argued in [13], the entropy of the superstar ensemble is maximised at vanishing chemical potential β. This is the regime in which we plan to compare the geometry (3.14) with the one obtained out of the proposal (3.54). First, the $\beta \to 0$ limiting behaviour of the finite chemical potential limit curve (3.35) describing the typical state of the superstar ensemble becomes a straight line

$$y = \frac{N_C}{N} x \equiv \omega\, x \,. \tag{3.55}$$

The grayscale distribution (3.53) will then be a constant fixing the superstar scalar function $z_S(0; r^2)$ to be

$$z_S(0; r^2) = \begin{cases} \frac{1}{2} \frac{N_C/N-1}{N_C/N+1} & \text{if } r^2/2\hbar \le N + N_C, \\ \frac{1}{2} & \text{if } r^2/2\hbar > N + N_C. \end{cases} \tag{3.56}$$

Since, within the droplet region, i.e. $r^2/2\hbar \le N + N_C$, this number is different from $\pm 1/2$, the spacetime is singular. Noting that the coarse grained phase space density derived from our proposal equals

$$u_S(0; r^2) = \frac{1}{2} - z_S(0; r^2) = \frac{1}{N_C/N + 1}, \tag{3.57}$$

in the region of the phase space plane between $r^2/2\hbar = 0$ and $r^2/2\hbar = N + N_C$, and vanishes otherwise, it is straightforward to check, using (3.46) and (3.47), that (3.56) reproduces (3.16).

Equation (3.56) is a prediction from our proposal and our gauge theory analysis concerning the description of the typical states in the semiclassical limit. We want to reproduce this prediction by explicit analysis of the configuration (3.14). If we compare the physical size of the two three-spheres appearing in the superstar metric (3.14) with their parametrisation in [105], we obtain the conditions:

$$\eta\, e^{G} = \sqrt{\gamma}\, r^2 \,, \qquad \eta\, e^{-G} = \frac{L^2 \sin^2 \theta_1}{\sqrt{\gamma}} \,. \tag{3.58}$$

Using the fact that $z = (1/2) \tanh G$ [105], then

$$z = \frac{1}{2} \frac{r^2\gamma - L^2 \sin^2 \theta_1}{r^2\gamma + L^2 \sin^2 \theta_1}. \tag{3.59}$$

Since it is the value $z(\eta = 0)$ that is related to the semiclassical distribution function, we must analyse the behaviour of G at $\eta = 0$. We observe there are two different regimes where this applies:

1. When $\sin\theta_1 = 0$, $z(\eta = 0) = 1/2$. Vanishing $\sin\theta_1$ implies the vanishing of the giant graviton distribution. Consequently, it should correspond to absence of fermion excitations in the gauge theory picture. This is precisely reflected in the boundary condition $z(\eta = 0) = 1/2$.
2. When $r = 0$, the giant distribution is non-vanishing. One gets

$$z(r = 0) = \frac{1}{2}\frac{\omega - 1}{\omega + 1}. \tag{3.60}$$

The gravity distribution (3.60) identically matches the one derived from purely field theoretic and statistical mechanical considerations (3.56). This establishes the singular superstar configuration corresponds to a coarse grained object matching the notion of typical state emerging from the canonical ensemble analysis.

3.3.4 Measurability, Coarse Graining and Entropy

The integrability of the system of N fermions and the non-renormalisation properties of this sector of the theory allow us to do better. First, we can be more precise about the kind of information loss occurring when taking the semiclassical limit and the subset of quantum states allowing a reliable gravitational description in that regime. Second, we can derive the semiclassical partition function from a first principle calculation on the gauge theory side that will highlight the definition of coarse graining as a renormalization group transformation in phase space.

To make integrability more explicit, one can either specify a basis of states in terms of the energies $\{f_1, \cdots f_N\}$ or in terms of the gauge invariant moments

$$M_k = \sum_{i=1}^{N} f_i^k = \mathrm{Tr}(H_N^k/\hbar^k) \; ; \quad k = 0, \cdots N, \tag{3.61}$$

where H_N is the N fermion Hamiltonian with the zero point energy removed. These M_k are conserved charges of the system of fermions in a harmonic potential [75] and allow a reconstruction of the spectrum $\mathcal{F} = \{f_1, \cdots f_N\}$ [14].

We are interested in reading this information from the bulk. Since the phase space distribution was identified with the scalar function $u(0; r^2)$, we search for the spectrum $\{M_1, \cdots M_N\}$ in its multipole expansion. After some algebra, one finds [14]

$$u(\rho, \theta) = 2\cos^2\theta \sum_{l=0}^{\infty} \frac{\hbar^{l+1}\sum_{f\in\mathcal{F}}A^l(f)}{\rho^{2l+2}}(-1)^l(l+1)\ _2F_1(-l, l+2, 1; \sin^2\theta), \tag{3.62}$$

where $_2F_1$ is the hypergeometric function, $A^l(f)$ is a polynomial of order l in f

$$A^n(f) \equiv \sum_{s=0}^{f} \frac{(-1)^{f-s}2^s f!}{(f-s)!s!}(s+1)_n , \qquad (3.63)$$

and $(\alpha)_n = \frac{(\alpha+n-1)!}{(\alpha-1)!}$ is the Pochhammer symbol. Thus, the data about the underlying state $\mathcal{F}$ enters the l-th moment in sums of the form

$$\sum_{f \in \mathcal{F}} A^l(f) = \sum_{k=0}^{l} c_k M_k \qquad (3.64)$$

where c_k is the coefficient of f^k in the polynomial expansion of $A^l(f)$. Thus a measurement of the first N multipole moments of the metric functions can be inverted to give the set of charges M_k of the underlying state, from which the complete wave function can be reconstructed.

The question is whether the above formal considerations survive the semiclassical limit. The latter consists of $\hbar \to 0$ with $\hbar N$ fixed. Moments M_k scale like $M_l = m_l N^{l+1}$. Hence, the multipole expansion reduces to [14]

$$u(\rho, \theta) = 2\cos^2 \theta \sum_{k=0}^{\infty} \frac{2^k \langle M_k \rangle}{\rho^{2k+2}} (-1)^k (k+1) \, _2F_1(-k, k+2, 1; \sin^2 \theta) . \quad (3.65)$$

Thus, there is a one–to–one correspondence between the bulk $u(\rho, \theta)$ multipole moments and the spectrum of the basis states encoded in the set M_k.

Since semiclassical observers have finite resolutions, we do not expect them to be able to measure all the required moments to identify the basis state. Indeed, to measure the l-th multipole in (3.65) one needs the (2l)-th derivative of the metric functions or any suitable invariant constructed from them. If the measuring device has finite size λ, the k-th derivative of a quantity within a region of size λ will probe scales of order λ/k. However, semiclassical devices can only measure quantities over distances larger than the Planck length. Thus,

$$\frac{\lambda}{k} > l_p = g_s^{1/4} l_s \qquad (3.66)$$

Setting the size λ to be a fixed multiple of the AdS scale $\lambda = \gamma L$, this says that

$$k < \gamma N^{1/4} \qquad (3.67)$$

for a derivative to be semiclassically measurable. Since we require $\mathcal{O}(N)$ multipoles to determine the quantum state and $N^{1/4}/N \to 0$ as $N \to \infty$, we conclude that semiclassical observers have access to a negligible fraction of the information needed to identify the quantum state. Reversely, it was shown in [14] that the distribution of low moments is universal, in the sense that the standard deviation to mean ratio of the moments M_k vanishes in the semiclassical limit. Thus, classical configurations have essentially identical low order multipoles, and their differences can not be observed.

These considerations are fairly generic, as argued in [15]. Assuming black holes are quantum mechanically described by a finite number of states, and consequently involve a discrete spectrum, it is clear that different quantum states can in principle be distinguished. The latter would imply there is no information loss. The catch is, once again, the necessary measurements required to tell these different quantum states apart involve Planck scale precision or waiting of the order of $\delta t \sim e^S$, due to the Heisenberg uncertainty principle.

So far we focused on the semiclassical measurability of states in the bulk, but nothing was explicitly mentioned about which subset of quantum states allows such a description. This is important because this process is intimately related to the emergence of entropy from the bulk perspective. To gain some insight into this issue, a second quantised formalism was developed in [16] to define an operator $\hat{u}(\alpha)$ whose expectation value in a generic state $|\Psi\rangle$ equals the one of the one-particle Husimi distribution[5]

$$\langle \Psi | \hat{u}(\alpha) | \Psi \rangle = \pi \mathrm{Hu}^{\rho_1}(\alpha) \,. \tag{3.68}$$

The operator in question is

$$\hat{u}(\alpha) \equiv b^{\dagger}(\alpha) b(\alpha) \,. \tag{3.69}$$

Here $b^{\dagger}(\alpha)$ stands for a fermion creation operator, whereas $|\alpha\rangle$ equals a coherent state localised in some point of phase space $\alpha = \frac{x_1 + i\,x_2}{\sqrt{2\hbar}}$,

$$|\alpha\rangle = e^{-|\alpha|^2/2} \sum_{n=0}^{\infty} \frac{\alpha^n}{\sqrt{n!}} |n\rangle \equiv \sum_{n=0}^{\infty} c_n(\alpha) |n\rangle \,. \tag{3.70}$$

Coherent states can be thought of states inhabiting a lattice of unit cell area $2\pi\hbar$ [5, 27, 133]. Since a semiclassical observer measures the phase plane at an area scale $\delta A = 2\pi\hbar M \gg 2\pi\hbar$, she is only sensitive to a smooth, coarse grained Wigner distribution $0 \le \hbar W_c = u_c \le 1$ erasing many details of the underlying precise microstates. The region δA consists of $M = \delta A / 2\pi\hbar$ lattice sites, a fraction $u_c = \hbar W_c$ of which are occupied by coherent states. Then the entropy of the local region δA is

---

[5]The exact quantum state is an N-particle state. Thus, there is generically information loss when going from this to the one particle description. In the large N limit, this is typically expected to be a subleading effect not emerging in the classical gravitational description. See [16] for a discussion on this matter.

$$\delta S = \log\left(\frac{M}{\hbar\, W_c\, M}\right) \sim -\frac{\delta A}{2\pi\hbar}\log u_c^{u_c}\,(1-u_c)^{1-u_c}\,, \tag{3.71}$$

when $\hbar W_c$ is reasonably far from 0 and 1. For the total entropy this gives [16]

$$S = \int \delta S = -\int dA\,\frac{u_c \log u_c + (1-u_c)\log(1-u_c)}{2\pi\hbar}\,. \tag{3.72}$$

Thinking about $u_c = \hbar W_c$ as the probability of occupation of a site by a coherent state, this is simply Shannon's formula for information in a probability distribution [144].[6] This procedure shows in a rather explicit way how entropy is generated by integrating out quantum data at smaller scales. The large amount of supersymmetry allows us to interpret this as gravitational entropy by providing a bridge connecting the gauge and gravity theory descriptions in the large N limit.

This picture allows us to compute the semiclassical partition function

$$Z = \int \mathcal{D}u(x_1, x_2)\,\mu(u)\,e^{-\beta(H(u)-\nu N(u))}\,, \tag{3.73}$$

where the measure $\mu(u)$ reflects not only the Jacobian in transforming between the supergravity fields and u, but also the number of underlying microscopic configurations that give rise to the same macroscopic spacetime. Our previous considerations suggest

$$\mu(u) = e^{-\int \frac{dx_1\,dx_2}{2\pi\hbar}(u\ln u + (1-u)\ln(1-u))} = e^{S(u)}\,, \tag{3.74}$$

where S is understood as the entropy of the spacetime. In the semiclassical limit, a spacetime is nonsingular if $u = 0, 1$ everywhere. In that case, $S(\mu) = 0$ and the measure μ is 1. In other words, semiclassical half-BPS spacetimes have an entropy if and only if they are singular.

Evaluating the partition function by the method of saddle point gives

$$\ln Z = \int \frac{d^2x}{2\pi\hbar}\ln(1 + e^{-\beta\frac{x_1^2+x_2^2}{2\hbar}+\beta\nu}) = \int_0^\infty ds\,\frac{s}{e^{s-\beta\nu}+1} \equiv \frac{1}{\beta}F_2(e^{\beta\nu})\,, \tag{3.75}$$

where $s = \beta(x_1^2 + x_2^2)/2$ and F_2 is a Fermi-Dirac function (Fig. 3.4).

This result can be derived from first principles by coarse graining the scale of the fundamental cells in the exact gauge theory partition function. This is defined as a renormalization group (RG) transformation in this space. Consider a lattice whose cells are $M \times M$ (in Planck units). From the microscopic point of view,

---

[6]This result was independently obtained by Masaki Shigemori in an unpublished work by considering a gas of fermionic particles in phase space.

Fig. 3.4 RG-transformation
in phase space giving rise to
entropy

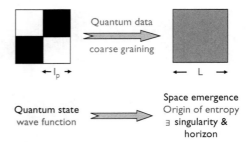

the energy of each distribution of populated Planck scale cells is different, but in the
limit $M \to \infty$, almost all distributions cluster close to a certain typical distribution
in the $M \times M$ cell, and thus observers at these scales will assign the same energy to
all of them. Another way to look into this transformation is more analogous to the
one taken in the semiclassical limit in gravity

$$\ell_p \to 0, \qquad L \to 0, \qquad \frac{L}{\ell_p} \to \infty \qquad (3.76)$$

L is the emergent continuous scale in the classical limit. It can be viewed as
$L = M\,\ell_p$. Before taking the limit, the coarse grained function u will take values
$0, \frac{1}{M^2}, \frac{2}{M^2}, \ldots, 1$ in the $M \times M$ cells. This can also be inferred by requiring the
phase space to describe N particles. Comparing the two lattices of sizes $M \times M$
and 1×1, in Planck units, one finds that

$$N = \sum_{\{x_1, x_2\}} u(x_1, x_2) = M^2 \sum_{\{x_1^M, x_2^M\}} u^M(x_1^M, x_2^M), \qquad (3.77)$$

allowing to derive that

$$u^M(x_1^M, x_2^M) = \frac{1}{M^2} \sum_{\substack{\{x_1, x_2\} \\ \in \{x_1^M, x_2^M\}}} u(x_1, x_2), \qquad (3.78)$$

where variables with superscript M are defined in the $M \times M$ lattice and in the
second equality we are summing over all Planck-scale lattice sites inside a single
$M \times M$ cell labelled by (x_1^M, x_2^M). This sum computes the fraction of populated
sites in the coarse grained cell. Finally, the sum over all possible u^M configurations
at cell location (x_1^M, x_2^M) becomes

$$\sum_{u^M} e^{-f\, u^M} = 1 + \binom{M^2}{1} e^{-f \frac{1}{M^2}} + \binom{M^2}{2} e^{-f \frac{2}{M^2}} + \ldots + e^{-f \frac{M^2}{M^2}} = \left(1 + e^{\frac{-f}{M^2}}\right)^{M^2}.$$
$$(3.79)$$

where $f(x_1, x_2) = \beta\left((x_1^2 + x_2^2)/2 - v\right)/(2\pi\hbar)$. The factors in front of each exponential count how many ways a given value of u^M in the coarse grained lattice can be attained in terms of the Planck scale lattice. These combinatorial factors equal the ones already argued for in the semiclassical considerations in (3.71). This is the precise origin of entropy when implementing the coarse graining transformation in the limit (3.76). The complete partition function becomes

$$Z^{M \times M} = \prod_{x_1, x_2 \in M\mathbb{Z}} \left(1 + e^{-\frac{\beta}{M^2}\left(\frac{x_1^2 + x_2^2}{2} - v\right)}\right)^{M^2} = \frac{M^2}{\beta} F_2(e^{\frac{\beta v}{M^2}}). \tag{3.80}$$

This derivation reproduces the semiclassical computation if we identify β in (3.75) with the rescaled potential β/M^2 due to the renormalization group transformation. One could interpret the computation (3.80) as a derivation for the entropy formula (3.71).

What this derivation highlights is that as soon as one observer has *no* access to Planck scale physics, the measured coarse grained phase space density will typically be fractional. The advantage of the system discussed above is that we can also explicitly see this in gravity in terms of singular configurations. This is a satisfactory explanation for the source of entropy in the semiclassical description.

3.3.5 Large Black Holes, Fuzzball and Ensembles

Testing the ideas behind the fuzzball conjecture has been reasonably successful for small supersymmetric black holes, the example discussed above being one particular example. See [17, 36, 120, 124, 146] for discussions and references involving other set-ups. It is natural to wonder how general and testable these ideas are for generic supersymmetric and non-supersymmetric black holes.

Even in the context of small supersymmetric black holes, it was pointed out in [143] that the existence of classical fuzzball configurations depends on the U-dual frame being considered. Even more, it was noticed there, that for all known examples, whenever such solutions do not exist, the small black hole develops a horizon through higher order corrections, whereas when it can be argued that the horizon scale remains of vanishing size, fuzzball configurations already exist at tree level. Sen conjectured this to be a generic fact [143]. This observation starts emphasising the importance of both α' and g_s corrections in these considerations, since both are not U-duality invariant.

What about large black holes? There exist candidate supersymmetric fuzzball configurations starting with the multi-center configurations in [31, 35, 43], including the scaling solutions [37, 74], where the coordinate distance between such centers goes to zero, and also more recent work [40] involving configurations with arbitrary functions. Their (classical) moduli spaces are much more complex than the ones for small black holes. Hence our understanding is far from complete (see [17] for a summary on the state of this important matter). But if fuzzball ideas were to be borrowed to these large black holes, one would ideally expect, at least in an euclidean path integral approach, that the partition function equals [146]

$$Z = \int_{f_a} D[g_a] \, D[\Phi_a] \, e^{-I_{f_a}} = e^{-I_{f_a}+S} = e^{-I_{BH}} \,, \tag{3.81}$$

where the sum is *only* carried over the fuzzball configurations f_a, with metric g_a, matter fields Φ_a and I_{BH} is the action for the black hole having the same mass and charges as the set $\{f_a\}$. Notice that the absence of a horizon for all f_a implies that the on-shell Euclidean action should satisfy

$$I_{f_a} = \beta \, (E - \mu_i \, Q_i) \,, \tag{3.82}$$

where the set $\{E, Q_i\}$ stands for the mass and charges carried by the black hole. Notice the above argument also used there is a total degeneracy $\mathcal{D} = e^S$ obtained after quantisation of the space of classical fuzzball solutions.

Using this formalism provides intuition, but its assumptions may not be generically fulfilled, especially in the semiclassical regime where they are technically applied. It is still useful to present some of the puzzles on the subject. The identity (3.81) assumes both the existence of solutions to the equations of motion of the effective action governing gravity and that their number explains the entropy of the original macroscopic black hole.

Let us put these assumptions in perspective with our previous discussions [145]. First, our knowledge on this effective action typically reduces to its tree level classical part. Thus, geometric quantisation deals with quantisation of a classical moduli space of configurations of a classical theory which is known to receive both α' and g_s corrections. Since the most conservative expectation is that information about individual microstates is stored in Planck scale physics, the use of the classical action is typically not justified to begin with. Thus, we expect *not* to be able to reproduce the entropy of the black hole from these considerations. Indeed, preliminary work in this direction [68, 69] confirms this expectation.

Second, there are matters of principle arising. Large lorentzian black holes typically have curvature singularities in their deep interior. Thus, they are *not* solutions to the classical equations of motion. Their *euclidean* continuations, however, are smooth when suitable boundary conditions are imposed, at the expense of removing the interior of the geometry. These euclidean configurations are *saddle points* of the semiclassical partition function and provide the dominant contributions to the macroscopic entropy d_{macro} by construction. But saddle points are believed *not* to provide a semiclassical description of states with well-defined mass and spin in a quantum gravity Hilbert space [77]. Thus, in an euclidean formalism, which is intrinsically canonical in nature, ensemble wise, the euclidean black hole definitely knows about the total number of states, but the information about the microstates seems to be lost (or it is not manifest in our current understanding).

The above remarks are consistent with the different available formulations of the AdS/CFT correspondence. The euclidean path integral allows to compute the partition function, and to extract the total number of states by saddle point approximation, but we also know that the use of lorentzian geometry is essential to capture the difference between microstates through the expectation values of the different gauge invariant operators encoded in the boundary fall-off conditions of the different bulk fields [8].

Furthermore, this discussion also highlights our ignorance on how to define path integrals more accurately. Indeed, in the examples were candidate fuzzball configurations are known, their euclidean continuations involve *complex* metrics, whereas their on-shell actions remain real.[7] Clearly, it is important to understand the space of euclidean configurations one needs to sum over. A different way of stressing this point is to notice that "thermal" circles in euclidean black holes are contractible, whereas the ones for known fuzzball configurations are not, suggesting topology can play a role in these considerations.

It is interesting to revisit our half-BPS sector of $\mathcal{N} = 4$ SYM discussion. This describes a small black hole in the bulk and it was argued in [145] that quantum corrections would not generate a macroscopic horizon. The large symmetry in this sector of the theory allowed us to match the semiclassical partition function with a first principle (quantum mechanical) derivation. The outcome of this matching was the derivation of the non-trivial measure $\mu(u)$ in (3.74). The correct partition function was obtained by summing over smooth and singular configurations in the semiclassical limit. If we would have ignored the non-trivial measure, the partition function would have reduced to

$$\ln \tilde{Z} = \int \frac{d^2 x}{2\pi} \ln \left(\frac{1 - e^{(-\beta \frac{x_1^2 + x_2^2}{2} + \beta v)}}{\beta \frac{x_1^2 + x_2^2}{2} - \beta v} \right) = \frac{1}{\beta} \int_0^\infty ds \ln \left(\frac{1 - e^{-s - \beta v}}{s - \beta v} \right). \quad (3.83)$$

There are two important points to be made. First, this integral diverges at the upper limit. Second, it would only reproduce (3.75) if one restricts the partition sum to be over smooth geometries ($u = 0, 1$) with u taking constant values within elementary cells at the Planck scale. Even though this would geometrically mimic the coherent state analysis in quantum mechanics, its validity is certainly doubtful in a manifest semiclassical treatment applicable all the way to the Planck scale.

It is interesting to point out the different role that different ensembles also play in Sen's independent approach to explain the entropy of supersymmetric extremal black holes with charges Q having AdS$_2$ throats from an entirely macroscopic perspective [141, 142]. Sen's proposal is

---

[7]This observation may not be that surprising since we know of examples in quantum mechanics in which the saddle point approximation involves a complex configuration. It seems still meaningful to appreciate its conceptual consequences beyond its purely technical nature.

$$d_{\text{macro}}(Q) = \sum_s \sum_{Q_i, Q_{\text{hair}}}^{\sum_{i=1}^{s} Q_i + Q_{\text{hair}} = Q} \left\{ \prod_{i=1}^{s} d_{\text{hor}}(Q_i) \right\} d_{\text{hair}}(Q_{\text{hair}}; \{Q_i\}) . \qquad (3.84)$$

The s-th term represents the contribution from an s-centered black hole configuration; $d_{\text{hor}}(Q_i)$ stands for the degeneracy associated with the horizon of the i-th black hole center carrying charge Q_i; and $d_{\text{hair}}(Q_{\text{hair}}; \{Q_i\})$ stands for the hair degeneracy, i.e. smooth black hole deformations supported outside the horizon and sharing the same asymptotics.

Sen's prescription uses a mixture of formulations. Indeed, whereas the contribution from the degrees of freedom localised at the horizon is captured by an euclidean path integral, both the contribution from horizonless configurations, through geometric quantisation, and hair modes employ entirely lorentzian methods. At any rate, a better understanding on how to formulate gravitational path integrals more rigorously is clearly desirable from many points of view.

Even though the above arguments strongly suggest the fuzzball programme should not work at tree level for non-supersymmetric configurations, this does not forbid, a priori, the existence of *non-typical* non-extremal fuzzball configurations solving the classical equations of motion. The first known examples of these were found in [101]. Remarkably, there exists an interesting body of work for some of these non-extremal configurations giving evidence that some features of these fuzzball ideas are still realised in less symmetric situations [58]. More recently, there has also been some progress in finding non-extremal fuzzball like configurations [38, 39] and explicit multi-center extremal non-BPS solutions (see [67] and references there in).

3.4 Extremal Black Holes

In the seminal work of Strominger and Vafa [149] the entropy of a certain supersymmetric black hole was accounted for by identifying its degrees of freedom with those of a 2d CFT. Later, it was realised [147] that matching the universality of the 2d CFT Cardy's formula (3.3) with the universality of the Bekenstein-Hawking entropy formula (3.2) follows from the seminal work of Brown and Henneaux [50] analysing the asymptotic symmetry group in the AdS$_3$ region emerging near the horizon of the original black hole.

The above synthesises the two most common approaches used in the string theory community to explain the macroscopic entropy of a given black hole. The first one is microscopic in nature. One maps black hole charges to D-brane charges (possibly) wrapping some internal cycles in some compact manifold. These provide an open string (gauge theory) description in which one counts the number of microstates compatible with the given conserved charges. The matching with the gravitational macroscopic result relies on the existence of non-renormalisation theorems guaranteeing the number of states does not change as the gauge coupling increases, which is what generates the gravity description. The second method is *semiclassical* in nature. It constructs a Hilbert space out of the study of the asymptotic symmetry group of a given spacetime. The emergence of a 2d conformal field theory realising two Virasoro algebras allows one to use Cardy's result (3.3) to account for the entropy of the black hole in terms of the number of operators carrying its charges.

The main goal of this section is to review part of the recent work devoted to extend the second approach to extremal, non necessarily BPS, black holes. More precisely, the Kerr/CFT correspondence [87] and its generalisation to extremal black holes/CFT [90].

3.4.1 Extremal Black Holes and Conformal Field Theory

Let me consider the semiclassical approach initiated by Brown and Henneaux in [50]. One of the virtues of this seminal work is to provide a semiclassical construction of a Hilbert space in a classical gravity theory given some boundary conditions. The heuristic idea is as follows. Given a reference metric g (global AdS$_3$ in [50]), one determines the subset of non-trivial diffeomorphisms ζ preserving some set of boundary conditions h (at infinity in [50])

$$\mathcal{L}_\zeta (g + h) \sim h \,, \tag{3.85}$$

where $\mathcal{L}_\zeta g$ stands for the Lie derivative of the metric g along the vector field ζ. These are understood as normalisable excitations of the background metric g. By non-trivial here, one means the associated conserved charge $Q_\zeta[g]$ does not vanish. One is interested in computing the algebra closed by these conserved charges under Dirac brackets, since the states in this semiclassical approximation will fit into representations of the latter. Thus one needs surface integrals defining them in terms of the given diffeomorphism ζ and the background metric g. Here, I follow the covariant formalism developed in [29, 30], based on [1] and further developed in [28, 59]. The charges generating ζ are

$$Q_\zeta = \frac{1}{8\pi G} \int_{\partial \Sigma} k_\zeta[h, g] \,, \tag{3.86}$$

where G is Newton's constant, $\partial \Sigma$ the boundary of a spatial slice and

$$k_\zeta[h, g] = \frac{1}{2} \Big[\zeta_\nu \nabla_\mu h - \zeta_\nu \nabla_\sigma h_\mu{}^\sigma + \zeta_\sigma \nabla_\nu h_\mu{}^\sigma + \frac{1}{2} h \nabla_\nu \zeta_\mu - h_\nu{}^\sigma \nabla_\sigma \zeta_\mu$$
$$+ \frac{1}{2} h_{\nu\sigma} (\nabla_\mu \zeta^\sigma + \nabla^\sigma \zeta_\mu) \Big] * (dx^\mu \wedge dx^\nu) \,, \tag{3.87}$$

All raised indices are computed using $g_{\mu\nu}$. The Dirac bracket algebra of the asymptotic symmetry group is then computed by varying the charges

$$\{Q_{\zeta_m}, Q_{\zeta_n}\}_{D.B.} = Q_{[\zeta_m,\zeta_n]} + \frac{1}{8\pi G} \int_{\partial\Sigma} k_{\zeta_m}[\mathcal{L}_{\zeta_n}g, g]. \tag{3.88}$$

Notice the resulting algebra can include a central term [29] if the last term does not vanish.

Recently, this philosophy was applied to extremal Kerr

$$ds^2 = -\frac{\Delta}{\rho^2}\left(d\hat{t} - a\sin^2\theta \, d\hat{\phi}\right)^2 + \frac{\sin^2\theta}{\rho^2}\left((r^2+a^2)\, d\hat{\phi} - a\, d\hat{t}\right)^2 + \frac{\rho^2}{\Delta}dr^2 + \rho^2\, d\theta^2, \tag{3.89}$$

where $\Delta = (r-a)^2$ and $\rho^2 = r^2 + a^2\cos^2\theta$. The Bekenstein-Hawking entropy equals

$$S = 2\pi J, \qquad \text{with} \qquad J = \frac{M^2}{G} \equiv G\, M_{ADM}^2. \tag{3.90}$$

Taking a near horizon limit [25]

$$t = \frac{\lambda \hat{t}}{2M}, \qquad y = \frac{\lambda M}{r-M}, \qquad \phi = \hat{\phi} - \frac{\hat{t}}{2M}, \qquad \lambda \to 0 \tag{3.91}$$

keeping (t, y, ϕ, θ) fixed, leads to the near-horizon extreme Kerr (NHEK) geometry

$$ds^2 = 2GJ\Omega^2\left(\frac{-dt^2 + dy^2}{y^2} + d\theta^2 + \Lambda^2\left(d\phi + \frac{dt}{y}\right)^2\right), \tag{3.92}$$

where $\Omega^2 \equiv (1 + \cos^2\theta)/2$ and $\Lambda \equiv 2\sin\theta/(1 + \cos^2\theta)$. This has an enhanced $SL(2, \mathbb{R}) \times U(1)$ isometry group acting on the fixed polar angle θ 3d slices, whose geometry is that of a quotient of warped AdS_3 and describes an S^1 bundle over an AdS_2 base.

To study the semiclassical excitations around NHEK, one studies its asymptotic symmetry group. To do that we impose the boundary conditions [87]

$$\begin{pmatrix} h_{\tau\tau} = \mathcal{O}(r^2) & h_{\tau\varphi} = \mathcal{O}(1) & h_{\tau\theta} = \mathcal{O}(r^{-1}) & h_{\tau r} = \mathcal{O}(r^{-2}) \\ h_{\varphi\tau} = h_{\tau\varphi} & h_{\varphi\varphi} = \mathcal{O}(1) & h_{\varphi\theta} = \mathcal{O}(r^{-1}) & h_{\varphi r} = \mathcal{O}(r^{-1}) \\ h_{\theta\tau} = h_{\tau\theta} & h_{\theta\varphi} = h_{\varphi\theta} & h_{\theta\theta} = \mathcal{O}(r^{-1}) & h_{\theta r} = \mathcal{O}(r^{-2}) \\ h_{r\tau} = h_{\tau r} & h_{r\varphi} = h_{\varphi r} & h_{r\theta} = h_{\theta r} & h_{rr} = \mathcal{O}(r^{-3}) \end{pmatrix}, \tag{3.93}$$

where the radial coordinate r in the global AdS_2 coordinates was used

$$y = \left(\cos \tau \sqrt{1 + r^2} + r \right)^{-1},$$

$$t = y \sin \tau \sqrt{1 + r^2}, \tag{3.94}$$

$$\phi = \varphi + \log \left(\frac{\cos \tau + r \sin \tau}{1 + \sin \tau \sqrt{1 + r^2}} \right).$$

The most general diffeomorphism preserving these boundary conditions is [87]

$$\zeta = \left(-r \epsilon'(\varphi) + \mathcal{O}(1) \right) \partial_r + \left(C + \mathcal{O}(r^{-3}) \right) \partial_\tau + \left(\epsilon(\varphi) + \mathcal{O}(r^{-2}) \right) \partial_\varphi + \mathcal{O}(r^{-1}) \partial_\theta,$$

where $\epsilon(\varphi)$ is an arbitrary smooth function of the periodic boundary coordinate φ and C is an arbitrary constant. Expanding $\epsilon(\varphi)$ into Fourier modes and defining dimensionless quantum versions of the Qs by $\hbar L_n \equiv Q_{\zeta_n} + \frac{3J}{2} \delta_n$ plus the usual rule of Dirac brackets to commutators as $\{.,.\}_{D.B.} \to -i/\hbar [.,.]$, the quantum charge algebra is then [60, 87]

$$[L_m, L_n] = (m - n) L_{m+n} + \frac{J}{\hbar} m(m^2 - 1) \delta_{m+n,0}. \tag{3.95}$$

This is a Virasoro algebra with central charge

$$c_L = \frac{12J}{\hbar}. \tag{3.96}$$

3.4.1.1 CFT Origin of the Gravitational Entropy

If the gravitational entropy of extremal Kerr allows a microscopic interpretation in terms of a chiral CFT as suggested by the chiral Virasoro algebra emerging from the previous semiclassical considerations, one is left to determine the temperature of the mixed state describing the NHEK geometry. To identify it, remember the state of an scalar quantum field in the Kerr background after integrating out its interior is given by a density matrix with eigenvalues

$$e^{-\hbar \frac{\omega - \Omega_H m}{T_H}}, \quad \text{with} \quad \Omega_H = \frac{a}{2M r_+}, \quad T_H = \frac{r_+ - M}{4\pi M r_+} \tag{3.97}$$

Here Ω_H is the angular velocity of the horizon and T_H is its Hawking temperature. We can relate these eigenvalues to the ones associated to the Killing vector fields ∂_ϕ and ∂_t naturally appearing in the near-horizon region through the identity

$$e^{-i\omega \hat{t} + im\hat{\phi}} = e^{-\frac{i}{\lambda}(2M\omega - m)t + im\phi} = e^{-in_R t + in_L \phi} \tag{3.98}$$

where

$$n_L \equiv m, \qquad n_R \equiv \frac{1}{\lambda}(2M\omega - m).$$ (3.99)

In terms of these variables the Boltzmann factor (3.97) is

$$e^{-\hbar\frac{\omega - \Omega_H m}{T_H}} = e^{-\frac{n_L}{T_L} - \frac{n_R}{T_R}},$$ (3.100)

where the dimensionless left right temperatures are

$$T_L = \frac{r_+ - M}{2\pi\,(r_+ - a)}, \qquad T_R = \frac{r_+ - M}{2\pi\lambda r_+}.$$ (3.101)

In the extremal limit $M^2 \to GJ$, these reduce to [87]

$$T_L = \frac{1}{2\pi}, \qquad T_R = 0.$$ (3.102)

The left-movers are then thermally populated with the Boltzmann distribution at temperature $1/2\pi$:

$$e^{-2\pi n_L},$$ (3.103)

while only the purely reflecting modes survive the limit since $\omega = m/(2M)$. Thus, even though extreme Kerr has zero Hawking temperature, the quantum fields outside the horizon are not in a pure state.

Assuming the existence of a unitary chiral CFT with central charge (3.96), one is tempted to appeal to Cardy's formula to account for the CFT entropy[8]

$$S_{\text{CFT}} = \frac{\pi^2}{3}c_L T_L.$$ (3.104)

Using (3.102) and (3.96), one reproduces the entropy of extremal Kerr (3.90) [87]

$$S_{\text{micro}} = \frac{2\pi J}{\hbar} = S_{BH}.$$ (3.105)

Notice this approach uses the symmetries emerging in the semiclassical analysis and the universality of Cardy's formula, but does not provide any explicit microscopic description of the system. This is a common feature of this kind of considerations.

This evidence was used in [87] to conjecture a new duality between quantum gravity in the near horizon of extremal Kerr and (a chiral half of) a two dimensional conformal field theory, the so called Kerr/CFT correspondence. By considering near-extremal Kerr, it was shown in [48] that the superradiant scattering of an scalar

---

[8]Cardy's formula requires the temperature to be large. See [100] for a justification on the validity of Cardy's regime for extremal Kerr.

field by a near-extreme Kerr black hole was fully reproduced by a two dimensional conformal field theory in which the black hole corresponds to a thermal state and the scalar field to a specific operator in the dual CFT, extending the standard AdS/CFT framework.

In the following, I briefly discuss how this same structure emerges for any extremal black hole giving rise to the extremal black hole/CFT conjecture [90, 107]. For preliminary work on the subject regarding the entropy of near-extremal black holes and the AdS$_2$/CFT$_1$ correspondence, see [130]. For a more complete set of references on the subject, see the recent review [49].

3.4.1.2 General Extremal Black Holes and Conformal Field Theory

Consider any asymptotically globally AdS (or Minkowski) extremal black hole solution to a general theory of $D = 4, 5$ Einstein gravity coupled to some arbitrary set of Maxwell fields F^I and neutral scalars ϕ^A. Extremality requires the existence of more than one charge besides mass. That is, either angular momentum, as in Kerr, or electric/magnetic charges. It was shown in [103] that assuming the black hole has a regular horizon and an $\mathbb{R} \times U(1)^{D-3}$ isometry group, Einstein's equations guarantee the corresponding near-horizon geometry is

$$ds^2 = \Gamma(\rho)\left[-r^2dt^2 + \frac{dr^2}{r^2}\right] + d\rho^2 + \gamma_{ij}(\rho)(dx^i + k^i rdt)(dx^j + k^j rdt)$$

$$F^I = d[e^I rdt + b^I_i(\rho)(dx^i + k^i_I rdt)]$$

$$\phi^A = \phi^A(\rho) \tag{3.106}$$

where $i = 1, \ldots D - 3, r = 0$ is the horizon, $\Gamma(\rho) > 0$, $\{\partial/\partial t, \partial/\partial x^i\}$ are Killing vector fields, and k^i, e are constants. The precise form of the ρ-dependent functions depends on the subset of field equations that have not yet been integrated. This metric has several S^1 bundles over AdS$_2$, the latter spanned by $\{t, r\}$. Thus, there exists an enhancement of symmetry to SO$(2, 1) \times$ U$(1)^{D-3}$. The presence of these bundles will be crucial to extend the previous semiclassical considerations leading to the existence of the chiral Virasoro algebra (3.95) for extremal Kerr. If the initial black hole only carries electric charges, this can be viewed as "rotation" using a convenient KK reduction from higher dimensions. Thus, the conclusions below would end up being the same [90].

To see how the ideas developed for extremal Kerr extend to more general situations, consider the most general near horizon geometry in 5d:

$$ds_5^2 = A(\theta)\left(-r^2dt^2 + \frac{dr^2}{r^2}\right) + F(\theta)d\theta^2 + B_1(\theta)\,\tilde{e}_1^2 + B_2(\theta)(\tilde{e}_2 + C(\theta)\,\tilde{e}_1)^2,$$

$$\tilde{e}_1 = d\phi_1 + k_1 r\,dt, \qquad \tilde{e}_2 = d\phi_2 + k_2 r\,dt, \tag{3.107}$$

where A, B_i, C and F are functions of the latitude coordinate θ (the analogue of ρ in (3.106)). The metric can be viewed as an S^3 bundle over AdS_2 and its Bekenstein–Hawking entropy is

$$S_{BH} = \frac{1}{4} \int d\theta \sqrt{B_1 B_2 F} \int d\phi_1 d\phi_2 . \tag{3.108}$$

It was shown in [57] that this near horizon geometry has a pair of commuting diffeomorphisms that generate two commuting Virasoro algebras

$$\zeta^1 = -e^{-in\phi_1} \frac{\partial}{\partial \phi_1} - in\, r\, e^{-in\phi_1} \frac{\partial}{\partial r} ,$$

$$\zeta^2 = -e^{-in\phi_2} \frac{\partial}{\partial \phi_2} - in\, r\, e^{-in\phi_2} \frac{\partial}{\partial r} . \tag{3.109}$$

Using the covariant formalism reviewed before, one can compute the central charges c_i in these Virasoro algebras [57]

$$c_i = \frac{3}{2\pi} k_i \int d\theta \sqrt{B_1 B_2 F} \int d\phi_1 d\phi_2 = \frac{6k_i S_{BH}}{\pi} , \quad i = 1, 2 \tag{3.110}$$

They reproduce the entropy of the original black hole, through Cardy's formula,

$$S_{BH} = \frac{\pi^2}{3} c_1 T_1 = \frac{\pi^2}{3} c_2 T_2 , \tag{3.111}$$

if the constants k_1 and k_2 are related to the CFT temperatures by

$$k_i = \frac{1}{2\pi T_i} . \tag{3.112}$$

It is reassuring to check that these are precisely the values that these constants take when we consider the near horizon of a given extremal black hole. Indeed, in that case,

$$T_i = \lim_{r_+ \to r_0} \frac{T_H}{\Omega_i^0 - \Omega_i} = -\frac{T_H'^0}{\Omega_i'^0} , \tag{3.113}$$

where Ω_i describe the angular velocities of the black hole at the horizon, T_H its Hawking temperature and quantities with a 0 label refer to their extremal values. These results can be generalised to higher dimensions [57].

These constants T_i appear naturally when expanding quantum fields in eigenmodes of the asymptotic charges. For an scalar field, after tracing over the interior of the black hole, the vacuum is a diagonal density matrix with eigenvalues

$$e^{-(\omega - \Omega_1 m_1 - \Omega_2 m_2)/T_H} . \tag{3.114}$$

Expanding $T_H = T'_H x$ and $\Omega_i = \Omega_i^0 + \Omega_i' x$, where x measures the distance to the extremal point, one concludes the density matrix after the extremal limit is given in terms of

$$e^{-\frac{m_1}{T_1} - \frac{m_2}{T_2}}, \tag{3.115}$$

with T_i defined as in (3.113). Thus these quantities can be interpreted as the Frolov–Thorne temperatures [82] associated with two CFTs, one for each azimuthal angle ϕ_i. Furthermore, the above requires the relation $\omega = \Omega_1^0 m_1 + \Omega_2^0 m_2$ among the different quantum numbers. These are the modes that are fully reflected from the black hole (no energy absorbed by the black hole) [25].

The existence of more than one CFT description may appear to be surprising. But we are well aware of this same fact for the black holes described in [149]. There, the entropy only depends on two of the three charges carried by the black holes, and depending on the U-duality frame being used, there are different available CFTs. In the current context, the existence of a lattice of CFTs was argued for in [106]. It is not clear whether the SL(2, $\mathbb{R}$) transformations acting on the moduli characterising the 5d near horizon geometry can be interpreted as a U-duality transformation. Embedding these systems in string theory, as in [100], could clarify this point.

The superradiant scattering of an scalar field by these backgrounds also matches the analogous calculation in a chiral 2d CFT, using the appropriate dual operator [65]. This provides similar evidence to the one reported for extremal Kerr. It is interesting to point out the work in [32], where besides providing further evidence for this correspondence, bulk correlators are computed for asymptotically flat black holes using the same recipe developed in the AdS/CFT correspondence, perhaps pointing towards the existence of new holographic relations for this different asymptotics.

Before closing this discussion, it is interesting to mention the potential connection between the results reviewed and further existent work in the literature. Prior to the Kerr/CFT conjecture, it was already observed that a quantum theory of gravity in 2d with negative cosmological constant coupled to an electric field could allow a non-trivial central charge under a suitable set of boundary conditions [89]. These were responsible for twisting the energy momentum tensor $T_{\pm\pm}$ generating 2d conformal transformations by the U(1) gauge current $j_\pm$ generating U(1) gauge transformations as

$$\tilde{T}_{\pm\pm} = T_{\pm\pm} \pm \alpha \, \partial_\pm j_\pm. \tag{3.116}$$

The constant α depends on the details of the theory (the AdS$_2$ radius ℓ and the electric field E in this case). Notice this twisting is responsible for generating a non-trivial central charge, since the original $T_{\pm\pm}$ has $c = 0$, as is customary in two spacetime dimensions. This mechanism was explicitly realised in a holographic formulation of AdS$_2$ black holes cross-checked from Kaluza-Klein compactification of the standard AdS$_3$/CFT$_2$ dictionary [54]. This may essentially be the same phenomena that is happening in the generic extremal near horizon geometry described in this section. If the starting extremal black hole has a compact horizon, its near horizon geometry (3.106) will allow, in principle, an effective description in terms of 2d gravity coupled to matter fields. One crucial property of the solutions to this theory is that all effective electric fields in 2d diverge at infinity. This is the reason why standard 2d conformal transformations must be accompanied by a U(1) gauge transformation to preserve the physical boundary conditions analysed in [89]. In other words, the singular behaviour of the electric field at infinity is the origin of the twisting. Interestingly, this twisting is reminiscent of the large gauge transformation that takes place in any near horizon limit for any extremal black hole. As one can see from (3.91), this limit involves two transformations

1. An IR limit, due to the red shift inherited by exploring the near horizon region of the starting extremal black hole, i.e. $r = r_h + \lambda\, y$ with $\lambda \to 0$
2. A large gauge transformation $t = \frac{\tau}{\lambda}$ and $\phi = \varphi + \Omega^0 \frac{\tau}{\lambda}$.

The large gauge transformation acts non-trivially on the generators of isometries ∂_t and ∂_ϕ on any quantum field propagating in this fixed background. At the level of the full theory, these symmetries are generated by the energy momentum tensor $T_{\pm\pm}$ and the U(1) gauge current $j_\pm$. Thus, the large gauge transformation implements the CFT twisting described in [89] at the level of the geometry. In the particular case where there is an available AdS/CFT description in the UV, one is tempted to view this near horizon limit as effectively implementing a non-trivial RG-flow in the field theory, by integrating out the spacetime outside of the black hole horizon. The non-trivial singular large gauge transformation required to keep the solution on-shell identifies the appropriate hamiltonian in the IR. For related discussions having condensed matter applications in mind, see the recent work [95].

3.4.1.3   Comments on the Existence of Non-trivial Dynamics

There are several arguments challenging whether the extremal BH/CFT correspondence has any dynamical content on it and just contains the degeneracy of the "vacuum" state

1. The existence of AdS$_2$ fragmentation [114] in the two-dimensional Einstein-Maxwell-Dilaton theory with a negative cosmological constant states that, at least classically, any matter excitation satisfying the null energy condition with support in the AdS$_2$ base will back-react strongly, modifying the spacetime boundary structure. This suggests there should be no states charged under the SL$(2, \mathbb{R})$ isometry group whenever this energy condition is satisfied and the physics we are interested in are described by the same effective 2d Einstein-Maxwell-Dilaton theory.
2. There are *no* normalisable linear perturbations of the NHEK geometry [2, 76] satisfying the boundary conditions described in [87]. In [2], it was further argued that the same conclusion holds for non-linear perturbations, which would match the above expectation in the semiclassical approximation.
3. Sen's work on extremal black holes reproducing microscopic results from a purely macroscopic point of view [138, 139], also concluded that when applying the straight AdS/CFT correspondence to the particular AdS$_2$/CFT$_1$

correspondence, the dual conformal quantum mechanics only includes the degeneracy of the vacuum, as also argued below [88, 140].

For asymptotically AdS black holes, this expectation can be understood as follows. Extremal black holes represent complicated mixed states in the dual UV CFT. Their excitations will have a gap if this CFT is non-singular and defined on the cylinder $\mathbb{R} \times S^{d-1}$. At sufficiently low energies above, less than the size of the gap, there will be no dynamics left, and no non-trivial theory remains. In the next subsection, we will see how this mechanism operates in AdS_3/CFT_2.

This argument already suggests a couple of ways to circumvent its conclusion

1. If the field theory lives on a *non-compact space*, its spectrum will be continuous. Thus the effective two dimensional Newton constant in AdS_2 will vanish, allowing us to bypass the fragmentation argument. This feature has appeared prominently in some recent applications of the AdS/CFT to condensed matter systems.[9]
2. Reduce the gap of the dual CFT by taking a *large central charge or large N limit*, since the gap typically scales with an inverse power of N or c. For finite size extremal black holes this would lead to a divergent entropy for a fixed temperature T. To obtain a finite entropy, it is tempting to consider large N limits together with a *vanishing horizon* limit.

Interestingly, it had previously been observed that under certain circumstances, whenever the horizon area of *extremal* black holes can be tuned to zero, their near-horizon geometries develop local AdS_3 throats [19, 25, 79, 80]. This is remarkable for at least two reasons

1. Given the AdS_3/CFT_2 correspondence, these particular points in the moduli space of extremal black holes may provide independent derivations for the existence of an IR CFT description of the black hole degrees of freedom.
2. Since these configurations are continuously connected to large extremal black holes where our previous considerations apply, one may identify the operator deforming the 2d CFT dual to AdS_3 and hope to be able to identify the finite deformation induced by it on the initial 2d CFT. This way, one could in principle try to derive whether the extremal black hole/CFT conjecture holds.

A general caveat about these classical configurations is its singular nature, and what their fate is when corrections are included in the bulk. There is an important distinction to be made in the cases discussed so far in the literature: when the system is near-extremal but far from BPS, the near horizon geometry involve a non-supersymmetric *pinching* $\mathbb{Z}_N$ orbifold of AdS_3,[10] whereas in the near BPS situation the transverse space decompactifies, but has an smooth 3d AdS throat.

---

[9]The amount of literature here is immense. We refer the reader to a subset of reviews and references therein [97].

[10]The action of this orbifold at the AdS_3 boundary is like the one of a conical defect. It would be interesting to see whether the techniques developed in [119] to compute the worldsheet string

Even though this direction will not be reviewed in these notes, it is worth mentioning two different approaches that have been followed

1. Given a near extremal black hole in AdS providing a well defined UV CFT dual description, one interprets the near horizon limit as a *large N IR limit* of the original CFT focusing in some low energy excitations of a definite sector of its Hilbert space selected by the *large gauge transformations* accompanying the near horizon limit. This is the approach followed in [19, 79, 80]. In this context, the conjectured CFT appearing in the extremal black hole/CFT correspondence emerges as an effective description for these excitations.

2. When no UV dual description is available, embedding the given black hole into string theory may provide with the existence of some points in the U-duality orbit where there exists a CFT dual. Tuning the charges of the black hole in that U-dual frame may allow the emergence of a local AdS_3 throat in the near horizon limit. One then identifies its central charge and temperature, matching the bulk entropy. Moving away from the point in charge space where the AdS_3 exists, one computes the linear deformation in the geometry allowing to identify the dual marginal operator deforming its dual 2d CFT dual. Ideally, the finite integration of this marginal deformation would connect the 2d CFT dual to AdS_3 to the one emerging in the extremal BH/CFT correspondence. This was the approach followed in [61, 86].

Despite these observations, one may still be interested in investigating whether the chiral CFT structure emerging in the strict extremal limit hides the existence of a non-chiral CFT as non-extremality is turned on. This is the direction pursued in [53, 55] for small non-extremality and in [32, 56] for finite non-extremality. The conclusion in all these works is affirmative, but further work is required to settle this very important question regarding non-extremal black holes. Both, research in extremal black hole microscopics and its applications to strongly coupled condensed matter systems suggest that the emergence of these CFTs appears at scales below a certain cut-off. Beyond it, extra degrees of freedom are necessary, and whether their interactions allow a CFT description remains an open question.

3.4.2 The AdS_3 Perspective

The physics of AdS_3 provides an excellent arena where to test the ideas previously described. First, AdS_3 allows a perturbative description as a 2d CFT string worldsheet [113]. Second, due to the lack of bulk degrees of freedom in three dimensions, black holes in 3d gravity with a negative cosmological constant correspond to quotients of global AdS_3 [23]. The latter also allow a perturbative worldsheet

---

perturbative spectrum can be extended to this case, and whether there is any interesting structure emerging in the large N limit.

description [115]. Third, due to the well-established AdS$_3$/CFT$_2$ correspondence, the system has a UV description in terms of a $1 + 1$ non-chiral CFT, prior to any low energy (near horizon) being considered. Finally, we already know that the two Virasoro algebras in this CFT are realised, in the semiclassical approximation, as a set of non-trivial diffeomorphisms preserving some set of boundary conditions defining the AdS$_3$ asymptotics [50].

In the following, our goal will be to understand the previous IR limits in terms of AdS$_3$/CFT$_2$. Let us first review both the bulk and CFT description of BTZ black holes [22]. These are asymptotically AdS$_3$ spacetimes with metric

$$ds^2 = -\frac{(r^2 - r_+^2)(r^2 - r_-^2)}{r^2\ell^2}dt^2 + \frac{\ell^2 r^2}{(r^2 - r_+^2)(r^2 - r_-^2)}dr^2 + r^2(d\phi - \frac{r_+ r_-}{\ell r^2}dt)^2 .$$
$$(3.117)$$

The periodicity $\phi \sim \phi + 2\pi$ makes them a quotient of AdS$_3$ [23]. Their ADM mass and angular momentum are

$$M\ell = \frac{r_+^2 + r_-^2}{8G_3\ell} , \quad J = \frac{r_+ r_-}{4G_3\ell} .$$
$$(3.118)$$

These depend on the radius of AdS$_3$ ℓ, 3d Newton's constant G_3 and both the inner and outer horizons, r_- and r_+, respectively.

Their dual interpretation is in terms of thermal states in a $1 + 1$ non-chiral CFT with left and right temperatures

$$T_L = \frac{r_+ + r_-}{2\pi\ell} , \qquad T_R = \frac{r_+ - r_-}{2\pi\ell} ,$$
$$(3.119)$$

related to the Hawking temperature of the BTZ black hole T_H by $\frac{2}{T_H} = \frac{1}{T_L} + \frac{1}{T_R}$.

The connection between the gravity and CFT descriptions is most easily reviewed following the asymptotic symmetry group analysis of Brown and Henneaux [50]. To do so, it is more convenient to work with lightlike coordinates $\hat{u} = t/\ell - \phi$ and $\hat{v} = t/\ell + \phi$, in which the global AdS$_3$ metric is $ds^2 = \ell^2(\frac{dr^2}{r^2} - 2r^2 d\hat{u}d\hat{v})$. The boundary conditions at large r are [50]

$$\delta g_{\hat{u}\hat{u}} \sim \delta g_{\hat{v}\hat{v}} \sim \delta g_{\hat{u}\hat{v}} \sim \mathcal{O}(1), \qquad \delta g_{rr} \sim \mathcal{O}\left(r^{-4}\right), \qquad \delta g_{r\hat{u}} \sim \delta g_{r\hat{v}} \sim \mathcal{O}\left(r^{-3}\right) .$$
$$(3.120)$$

That order one fluctuations in $\delta g_{\hat{u}\hat{u}}$, $\delta g_{\hat{v}\hat{v}}$ correspond to normalisable modes in the 2d CFT can be inferred by writing the BTZ black hole metric in these coordinates and examining its asymptotics. In this way, order $\mathcal{O}(1)$ fluctuations in $\delta g_{\hat{u}\hat{u}}$, $\delta g_{\hat{v}\hat{v}}$ are seen to independently change the mass and angular momentum in the dual $2d$ CFT.

The set of non-trivial charges constructed out of the diffeomorphisms preserving these boundary conditions close two commuting Virasoro algebras

$$[L_m, L_n] = (m - n)\, L_{m+n} + \frac{c}{12}m(m^2 - 1)\delta_{m+n,0}\,,$$

$$\left[\bar{L}_m, \bar{L}_n\right] = (m - n)\, \bar{L}_{m+n} + \frac{\bar{c}}{12}m(m^2 - 1)\delta_{m+n,0}\,, \tag{3.121}$$

with central charges

$$c = \bar{c} = \frac{3\ell}{2G_3}\,. \tag{3.122}$$

The generators L_n and $\bar{L}_n$ are given by

$$L_n - \frac{c}{24}\,\delta_{n,0} = e^{in\hat{v}}\,\frac{\partial}{\partial \hat{v}}\,, \qquad \bar{L}_n - \frac{\bar{c}}{24}\,\delta_{n,0} = e^{in\hat{u}}\,\frac{\partial}{\partial \hat{u}}\,, \tag{3.123}$$

whereas their zero modes are related to the bulk charges by

$$L_0 - \frac{c}{24} = \frac{M\ell + J}{2}\,, \qquad \bar{L}_0 - \frac{\bar{c}}{24} = \frac{M\ell - J}{2}\,. \tag{3.124}$$

The states in the Hilbert space thus constructed arrange themselves into representations of these Virasoro algebras. It is then a universal result that the number of highest weight operators/states carrying conformal dimension $\Delta = L_0 + \bar{L}_0$ and spin $L_0 - \bar{L}_0$ is given by Cardy's formula (3.3) [52]. Using the dictionary described above

$$S_{\text{Cardy}} = 2\pi\sqrt{\frac{c}{6}\left(L_0 - \frac{c}{24}\right)} + 2\pi\sqrt{\frac{\bar{c}}{6}\left(\bar{L}_0 - \frac{\bar{c}}{24}\right)} = \frac{2\pi r_+}{4G_3} = S_{\text{B-H}}\,, \tag{3.125}$$

one always reproduces the Bekenstein-Hawking formula (3.2) using the BTZ metric (3.117).

3.4.2.1   Near Horizon of Extremal BTZ Black Holes

Consider the subset of extremal BTZ black holes for which the inner horizon coincides with the outer one. Denote the horizon by $r_h = r_+ = r_-$, then

$$ds^2 = -\frac{(r^2 - r_h^2)^2}{r^2\,\ell^2}\,dt^2 + \frac{\ell^2\,r^2}{(r^2 - r_h^2)^2}\,dr^2 + r^2\left(d\phi - \frac{r_h^2}{r^2}\frac{dt}{\ell}\right)^2\,. \tag{3.126}$$

The UV dual description of this limit, $M\ell = J$, involves setting the right-movers to their ground state

$$\bar{L}_0 = \frac{c}{24}\,; \quad T_R = 0\,, \tag{3.127}$$

while the left moving temperature $T_L = \frac{r_h}{\pi\ell}$ and L_0 remain arbitrary.

Following the philosophy discussed for generic extremal black holes in four and five dimensions, we want to recover the same physics from the low energy limit involved when taking the near horizon limit of these black holes. To study the latter, introduce new coordinates

$$\hat{u} = t/\ell - \phi, \qquad \hat{v} = t/\ell + \phi, \qquad r^2 - r_h^2 = \ell^2 e^{2\rho}, \tag{3.128}$$

in which the metric (3.126) takes the form

$$ds^2 = r_h^2 \, d\hat{u}^2 + \ell^2 \, d\rho^2 - \ell^2 e^{2\rho} \, d\hat{u} \, d\hat{v}. \tag{3.129}$$

The near horizon limit consists of taking $\rho_0 \to -\infty$

$$\rho = \rho_0 + r, \quad u = \hat{u} \frac{r_h}{\ell}, \quad v = e^{2\rho_0} \frac{\ell}{r_h} \hat{v}, \quad \{u, v\} \sim \{u - 2\pi \frac{r_h}{\ell}, v + 2\pi \frac{\ell}{r_h} e^{2\rho_0}\} \tag{3.130}$$

while keeping r, u, v and r_h fixed. The resulting near horizon metric

$$ds^2 = \ell^2 (du^2 + dr^2 - e^{2r} \, du \, dv) \tag{3.131}$$

is locally identical to (3.129) but the boundary periodicities in the limit are

$$\{u, v\} \sim \{u - 2\pi \frac{r_h}{\ell}, v\}. \tag{3.132}$$

Thus, the boundary of (3.131) ($r \to \infty$) is a "null cylinder" – it has a metric conformal to $du\,dv$, the standard lightcone metric on a cylinder, but has a single compact *null* direction (u).

The metric (3.131) is the spacelike self-dual orbifold [10, 63],[11] an S^1 fibration over an AdS$_2$ base with isometry group SL$(2, \mathbb{R}) \times$ U(1), which is more easily seen when written as

$$ds^2 = \frac{\ell^2}{4} \frac{d\rho^2}{\rho^2} - \frac{\rho^2}{r_h^2} \frac{d\tau^2}{\ell^2} + r_h^2 \left(d\varphi + \frac{\rho}{r_h^2} \frac{d\tau}{\ell}\right)^2 = \frac{\ell_3^2}{4} \frac{d\rho^2}{\rho^2} + 2\frac{\rho}{\ell} d\tau d\varphi + r_h^2 d\varphi^2. \tag{3.133}$$

To understand the physical meaning of a null boundary cylinder, introduce a UV cut-off in the dual CFT by considering bulk surfaces of fixed large r

$$ds^2 = du^2 - e^{2r} \, du \, dv. \tag{3.134}$$

It was shown in [10] that (3.134) is conformal to a boosted cylinder. As $r \to \infty$ the boost becomes infinite, precisely matching the procedure defined by Seiberg [137]

---

[11] The importance of this geometry for the physics of extremal black holes was already emphasised some time ago in [10, 148].

for realising the Discrete Light Cone Quantization (DLCQ) of a field theory.[12] This suggests the spacelike self-dual orbifold is dual to the DLCQ of the original $1 + 1$ non-chiral CFT. We will see later the latter only keeps one chiral sector of the original UV CFT, in agreement with our previous Kerr/CFT considerations. The only parameter of the spacelike self-dual orbifold metric, r_h, is then related to the value of the light-cone momentum p^+ defining the DLCQ sector

$$p^+ = \frac{c}{6} \left(\frac{r_h}{\ell}\right)^2 . \tag{3.135}$$

This conclusion can also be reached by computing its boundary stress-tensor [21]. The spacelike self-dual orbifold (3.133) has a finite temperature [20][13]

$$T_{\text{self-dual}} = \frac{r_h}{\pi \ell} = \sqrt{\frac{6p^+}{\pi^2 c}} , \tag{3.136}$$

agreeing with T_L in the UV description. This same temperature could have been derived using the general discussion in Sect. 3.4.1.2. In particular, using the relation (3.112) between the CFT temperatures T_i and the constants k_i appearing in the near horizon extremal geometries using the same normalisation as in (3.106).

The $r_h \to 0$ limit of the spacelike self-dual orbifold sends the temperature (3.136) to zero and yields the metric

$$ds^2 = r^2 dx^+ dx^- + \ell^2 \frac{dr^2}{r^2} , \qquad x^- \sim x^- + 2\pi , \tag{3.137}$$

where we have conveniently renamed $\rho = r^2$, $\varphi = x^-$ and $\tau = 2\ell x^+$. The causal character of the compact direction x^- has changed, from an everywhere spacelike direction (except at the boundary) to an everywhere null direction. Thus, (3.137) should be identified with the null self-dual orbifold.[14] By construction, this spacetime contains closed lightlike curves but it has the same boundary as (3.133). Thus, it can be viewed as a different state belonging to the same DLCQ CFT.

---

[12]The precise definition of DLCQ in quantum field theory is rather subtle. As emphasised in [96], amplitudes computed in these theories diverge order by order in perturbation theory due to strong interactions among longitudinal zero modes. This quantization scheme was argued to be well defined non-perturbatively.

[13]There are different ways of arguing the existence of this temperature. From the global version of the spacelike self-dual orbifold [10, 21] containing two disjoint causally connected boundaries, the finite temperature originates from entanglement entropy after integrating out part of the space leading to the single boundary metric (3.133), pretty much in the same way Rindler space has a finite temperature when viewed as a local patch of the full Minkowski spacetime.

[14]Readers interested in the supersymmetric properties of this orbifold, see [63, 81]. In particular, [81] discusses the embedding of this orbifold in higher dimensional supergravities stressing the importance of the fermion chirality to assess the supersymmetry of this quotient.

Since the $r_h \to 0$ limit corresponds to $p^+ \to 0$, the null self-dual orbifold should correspond to the $p^+ = 0$ sector of the DLCQ CFT.

Since taking $r_h \to 0$ in (3.126) corresponds to the massless BTZ black hole

$$ds^2 = r^2 d\tilde{x}^+ d\tilde{x}^- + \ell^2 \frac{dr^2}{r^2} \qquad \tilde{x}^\pm = \phi \pm t/\ell , \qquad \phi \sim \phi + 2\pi , \qquad (3.138)$$

it is natural to view the null self-dual orbifold as its near horizon geometry. Indeed, consider

$$r = \epsilon \rho , \qquad \tilde{x}^- = x^- , \qquad \tilde{x}^+ = \frac{x^+}{\epsilon^2} , \qquad \epsilon \to 0 . \qquad (3.139)$$

The lightlike direction $\tilde{x}^+$ effectively decompactifies, while x^- remains compact $x^- \sim x^- + 2\pi$. Thus, the near horizon limit (3.139) of a massless BTZ black hole is the null self-dual AdS$_3$ orbifold (3.137).

3.4.2.2 Exciting the Null Self-Dual Orbifold

If our interpretation is correct, the spacelike self-dual orbifold (3.133) should be viewed as an excitation over the null self-dual orbifold (3.137). In particular, injecting some chiral momentum into the system keeping its causal null cylinder boundary should correspond to the spacelike self-dual orbifold. This is achieved by adding some wave to the conformally flat metric

$$ds^2 = \frac{\ell^2}{z^2} \left[dx^+ dx^- + k z^2 (dx^-)^2 + dz^2 \right] . \qquad (3.140)$$

Since there are no propagating degrees of freedom in d = 3, the latter is locally AdS$_3$, and it is indeed isometric to the spacelike self-dual orbifold (3.133), with r_h^2 being replaced with $k\ell^2$.

All these observations are consistent with the well-known fact that extremal BTZ is a chiral excitation above the massless BTZ black hole [47, 66]. This is the three dimensional counterpart of the easiest constructions of non-relativistic gravity duals to DLCQ CFTs in higher dimensions [83, 116], the main difference here being the non-dynamical character of 3d gravity. Notice the only non-singular non-relativistic gravity dual corresponds to the sector of large p^+, as is customary in gauge/gravity theory dualities and Matrix theory [7, 24].

3.4.2.3 The Pinching $\mathbb{Z}_N$ Orbifold

There exists a second inequivalent near horizon limit one could take from (3.138)

$$r = \epsilon \rho , \qquad \hat{x}^\pm = \epsilon x^\pm . \qquad (3.141)$$

Fig. 3.5 Relation between
different local AdS$_3$
geometries, their near horizon
limits and their dual
interpretations

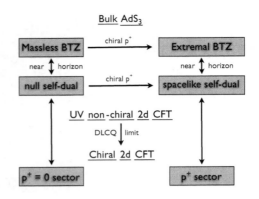

The resulting geometry is locally AdS$_3$

$$ds^2 = \rho^2 d\hat{x}^+ d\hat{x}^- + \ell^2 \frac{d\rho^2}{\rho^2} \qquad \hat{x}^\pm \sim \hat{x}^\pm + 2\pi\epsilon\,. \tag{3.142}$$

I will refer to it as a pinching AdS$_3$ orbifold [70] given its action on the boundary, which becomes a *pinching cylinder*, $R \times S^1/\mathbb{Z}_N$, with $N = 1/\epsilon$. In the bulk though, the quotient is that of a massless BTZ with periodicity scaling to zero.[15] One way of getting some intuition for what this may mean is to consider the identity [70]

$$\text{BTZ}(M\lambda^2, J\lambda^2; 2\pi) \equiv \text{BTZ}(M, J; 2\pi\lambda)\,, \tag{3.143}$$

This states that a BTZ black hole with pinching periodicity $2\pi\lambda$ is classically equivalent to a BTZ black hole with standard periodicity but mass and angular momentum scaled by λ^2. At any rate, all these geometries are singular, and a proper understanding of the physics for these values of the parameters requires to go beyond the classical gravitational approximation. Since the pinching orbifold is *not* equivalent to the null self-dual orbifold, this establishes that different near horizon limits can indeed capture different physics, as we will argue below (Fig. 3.5).

3.4.2.4 Low Energy IR Limits of 2d Non-chiral CFTs

Let us interpret the bulk near horizon limits in the dual CFT theory. Consider a non-singular non-chiral $2d$ CFT on a cylinder of radius R

$$ds^2 = -dt^2 + d\phi^2 = -du'\,dv' \quad ; \quad u' = t - \phi,\ v' = t + \phi\,. \tag{3.144}$$

---

[15]The same structure appears in the near horizon of extremal black holes with vanishing horizon. See [4] for different examples of its appearance.

Since $\phi \sim \phi + 2\pi R$, the light-like coordinates satisfy $\{u', v'\} \sim \{u' - 2\pi R, v' + 2\pi R\}$. Let $P^{u'}$ and $P^{v'}$ denote momentum operators in the v' and u' directions respectively. Their eigenvalues

$$P^{v'} = \left(h + n - \frac{c}{24}\right)\frac{1}{R}, \qquad P^{u'} = \left(h - \frac{c}{24}\right)\frac{1}{R}, \qquad n \in \mathbb{Z} \qquad (3.145)$$

are given in terms of the quantised momentum n along the S^1, the $2d$ central charge c and an arbitrary value of h with $h \geq 0$ and $h + n \geq 0$ due to unitarity constraints. These are related to the eigenvalues of the standard operators $L_0, \bar{L}_0$ used in radial quantisation on the plane by $\bar{L}_0 = h + n$ and $L_0 = h$.

Let us first show the DLCQ of this 2d non-chiral CFT is a 2d chiral CFT. Following Seiberg [137], we will study the consequences of the kinematics of an infinite boost on the discrete spectrum of the theory. We do this because the boundary structure of the near horizon bulk geometry was interpreted above as an infinitely boosted cylinder. Consider a boost with rapidity γ and take the double scaling limit

$$u' \rightarrow e^{\gamma} u', \qquad v' \rightarrow e^{-\gamma} v', \qquad \gamma \rightarrow \infty, \qquad R_- \equiv R e^{\gamma} \text{ fixed} \qquad (3.146)$$

The metric is invariant but the cylinder periodicities become

$$\begin{pmatrix} \phi \\ t \end{pmatrix} \sim \begin{pmatrix} \phi \\ t \end{pmatrix} + \begin{pmatrix} 2\pi R \\ 0 \end{pmatrix} - \text{infinite boost} \rightarrow \begin{pmatrix} u' \\ v' \end{pmatrix} \sim \begin{pmatrix} u' \\ v' \end{pmatrix} + \begin{pmatrix} -2\pi R_- \\ 2\pi R_- e^{-2\gamma} \end{pmatrix}$$

$$(3.147)$$

We can now identify $\{u', v'\}$ with the light-like boundary coordinates of AdS$_3$ in (3.130) via $u' = u(\ell/r_+)R_-$ and $v' = v(r_+/\ell)R_-$. Comparing (3.130) and (3.147), we conclude that the action of the near horizon limit on u, v precisely reproduces the identifications induced by the double scaling limit (3.146). Thus, the dual to the near-horizon geometry of the extremal BTZ black hole should be the DLCQ of the $1+1$ dimensional CFT dual to AdS$_3$.

Let us study the states that survive the double scaling limit (3.146). First, because of the kinematics of the DLCQ boosts,

$$P^{v'} = \left(h + n - \frac{c}{24}\right)\frac{e^{-\gamma}}{R}, \qquad P^{u'} = \left(h - \frac{c}{24}\right)\frac{e^{\gamma}}{R}. \qquad (3.148)$$

Keeping $P^{u'}$ finite in the $\gamma \rightarrow \infty$ limit requires $h = c/24$. This leads to

$$P^{v'} = n \cdot \frac{e^{-\gamma}}{R} = \frac{n}{R_-}. \qquad (3.149)$$

The DLCQ limit generates an infinite energy gap in the right-moving sector. Thus, keeping only finite energy excitations, it freezes to its ground state. The energy gap in the left-moving sector is kept finite. All physical finite energy states only carry momentum along the compact null direction u'. Therefore, the Hilbert space $\mathcal{H}$ of the DLCQ of the original 2d non-chiral CFT is

$$\mathcal{H} = \{|\text{anything}\rangle_L \otimes |c/24\rangle_R\}. \tag{3.150}$$

The chirality of the DLCQ theory spectrum can also be seen by studying which subset of the original AdS$_3$ Virasoro generators (3.121) remains under the double scaling limit (3.146) [20].

The null self-dual orbifold is now easily interpreted. Since it has the same boundary structure as the spacelike self-dual orbifold, it also corresponds to a state in the DLCQ theory. But, it describes its vanishing momentum $p^+ = 0$ sector.

The physical interpretation of the pinching $\mathbb{Z}_N$ orbifold must be different. Since the latter corresponds to sending the radius R of the limiting boundary cylinder to zero, $R \sim \epsilon \to 0$, it certainly generates an infinite gap in the untwisted sector for both chiral sectors of the initial 2d CFT. The only surviving untwisted finite excitations are those corresponding to $h = n = 0$. Thus, given a CFT with a fixed central charge c, this near horizon limit freezes out both left and right moving sectors, leaving us with the Hilbert space:

$$\mathcal{H} = \{|c/24\rangle_L \otimes |c/24\rangle_R\}. \tag{3.151}$$

But this simple argument does, a priori, not capture the full perturbative string spectrum. It would be interesting to extend the results in [119] for this case, clarifying whether there exists any massless twisted modes and whether their dynamics simplifies in the $N \to \infty$ limit.

3.4.2.5 Asymptotic Symmetries and the Chiral Virasoro Algebra

Our arguments above suggest that half of the available UV Virasoro generators become irrelevant for the IR physics captured by the near horizon extremal geometry. One way of checking this would be to study how these generators (3.123) transform under the infinite Lorentz boost defining the DLCQ limit of the original 2d non-chiral CFT. Instead, one can study the asymptotic symmetry group preserving the near horizon geometry.

The problem is then to identify the subset of non-trivial diffeomorphisms preserving the boundary conditions defining an asymptotically spacelike self-dual orbifold. Since these spaces are locally AdS$_3$, the analysis must be very similar to the one in [50]. The main physical insight comes from the observation that a general deformation of $\delta g_{\hat{u}\hat{u}}$, $\delta g_{\hat{v}\hat{v}}$ would be non-extremal and would thus excite both chiral sectors of the UV dual CFT. By contrast, we want to restrict to extremal excitations. Imposing the extremality condition $L_0 = c/24$ requires a more stringent boundary condition on the variations in $g_{\hat{v}\hat{v}}$. In [20], it was suggested to replace the boundary condition on $g_{\hat{v}\hat{v}}$ by

$$\delta g_{\hat{v}\hat{v}} \sim \mathcal{O}(r^{-2}). \tag{3.152}$$

The connection to the Brown-Henneaux diffeomorphisms is now made explicit: the diffeomorphisms generated by $\zeta = \zeta^\alpha \partial_\alpha$ are exactly of the form

$$\zeta^u = 2f(u) + \frac{1}{2r^2}g''(v) + \mathcal{O}(r^{-4}) \tag{3.153a}$$

$$\zeta^v = 2g(v) + \frac{1}{2r^2}f''(u) + \mathcal{O}(r^{-4}), \tag{3.153b}$$

$$\zeta^r = -r\left(f'(u) + g'(v)\right) + \mathcal{O}(r^{-1}) \tag{3.153c}$$

They satisfy the constraint

$$g'''(v) = 0 \quad \Longrightarrow \quad g(v) = A + B\,v + C\,v^2 . \tag{3.154}$$

implementing the boundary condition (3.152). Thus, one set of allowed diffeomorphisms is specified by a periodic function $f(u) = f(u + 2\pi)$. The analysis of generators of these diffeomorphisms follows directly from those of Brown and Henneaux, leading to a *chiral Virasoro algebra* at central charge $c = 3\ell/2G_3$. The remaining three parameter family of diffeomorphisms in (3.154) describes the $SL(2, \mathbb{R})$ isometries of the spacelike self-dual orbifold and act trivially on the Hilbert space [20].

Notice this construction mimics the phenomena reported for extremal Kerr and for general extremal black holes in $d = 4,5$ dimensions described in Sect. 3.4.1.2. Indeed, one starts with an extremal BTZ black hole, whose near horizon geometry consists of an S^1 fibration over AdS_2. The latter has isometry group $SL(2, \mathbb{R}) \times U(1)$. This gets enlarged to a full chiral Virasoro, but the infinite asymptotic symmetry algebra extends its $U(1)$ global part, whereas the $SL(2, \mathbb{R})$ acts trivially.

The virtue of the AdS_3 set-up is the existence of a well-defined UV dual CFT description allowing us to interpret the near horizon limit as an IR limit that turns out to be equivalent to a DLCQ limit. This gives some validity to the general arguments given in previous sections, but it does not clarify whether the Kerr/CFT correspondence is correct. Indeed, one of the original motivations in [20] was to argue that the mechanism behind the Kerr/CFT conjecture was three dimensional in nature. Some further supporting evidence was given in [55], where it was explicitly checked that the twisted CFT advocated in [89] was consistent with an AdS_3 reduction to AdS_2.[16]

3.4.2.6 Large N Limits, Double Scaling Limits and Existence of Dynamics

The double scaling limit discussed in Sect. 3.4.1.3 is easy to implement in the AdS_3 context [70]. In gravity, one is forced to consider a vanishing horizon limit, $r_\pm \to \epsilon r_\pm$ with $\epsilon \to 0$, while scaling Newton's constant $G_3 \to \epsilon G_3$, to keep the full

---

[16]For a different emphasis on how to use the AdS_3/CFT_2 correspondence to learn how to formulate the AdS_2/CFT_1 correspondence, see [88].

entropy finite. One can achieve this on BTZ metrics by combining a near horizon limit with this double scaling limit

$$r_\pm = \epsilon \rho_\pm \,, \quad r = \epsilon \rho_+ + \epsilon \rho \,, \quad t = \epsilon^{-1} \tau \,, \quad \phi = \epsilon^{-1} \psi \,, \quad G_3 = \epsilon \tilde{G}_3 \quad \epsilon \to 0 \,. \tag{3.155}$$

This reproduces the double scaling limit one could have considered in terms of 2d CFT data. Indeed, the latter corresponds to $R \to 0$, $c \to \infty$, $cR =$ fixed, where R stands for the radius of the original cylinder. Notice the scaling of R is related to the presence of a pinching orbifold in the gravity construction. Both transformations achieve $c \to c/\epsilon$, $L_0 - \frac{c}{24} \to \epsilon(L_0 - \frac{c}{24})$ and $\bar{L}_0 - \frac{c}{24} \to \epsilon(\bar{L}_0 - \frac{c}{24})$ as required.

One can gain some intuition about the different 2d CFT's appearing in this discussion by thinking about the CFT dual to the D1–D5 system. This 2d CFT can be described by a 2d sigma model with $N = (4, 4)$ supersymmetry with a target which can be thought of as a suitable symmetric product $\mathrm{Sym}^{N_1 N_5}(\mathcal{M}_4)$. Rescaling the central charge is like rescaling $N_1 N_5$. Therefore, one expects a relation of the form

$$\mathrm{CFT}_{\mathrm{new}} \approx \mathrm{Sym}^K(\mathrm{CFT}_{\mathrm{old}}) \,. \tag{3.156}$$

This clearly only makes sense when $K = 1/\epsilon$ is an integer. The new CFT has a long string sector which is directly inherited from the old theory. Its Virasoro algebra is related to that of the original CFT using standard orbifold technology. Explicitly, given a set of generators $\{L_n\}$, consider the subalgebra with generators[17]

$$l_n \equiv \frac{1}{K} L_{nK} \,, \quad n \neq 0 \,, \qquad l_0 \equiv \frac{1}{K}(L_0 - \frac{c}{24}) + \frac{c}{24} K \,. \tag{3.157}$$

It is then straightforward to see that l_n also form a Virasoro algebra with central charge $c' = cK$ and that the spectrum of l_0 has a spacing of $1/K$ compared to that of L_0.

Since the long string sector in an orbifold theory tends to dominate the entropy, this provides a natural explanation for the constancy of the entropy in the bulk. Further work is required to clarify the fate of these constructions given the $K \to \infty$ nature of the limit and the role twisted states must play in the resolution of the singular classical geometries corresponding to the nearly massless BTZ and its near horizon geometries involved in these limits.

---

[17]Notice the transformation for the generator l_0 is due to the fact that we were working on the plane. Indeed, if we would have worked on the cylinder, the transformation is the expected one:

$$l_n^{\mathrm{cyl}} \equiv \frac{1}{K} L_{nK}^{\mathrm{cyl}} \,, \quad n \neq 0 \,, \qquad l_0^{\mathrm{cyl}} \equiv \frac{1}{K} L_0^{\mathrm{cyl}} \,.$$

We now see that the transformation quoted on the plane makes sure the above cylinder transformation brings us back to the plane.

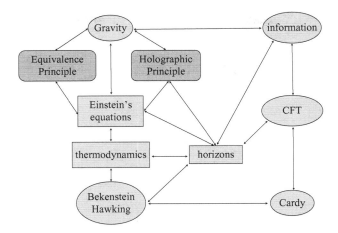

Fig. 3.6 Different perspectives on gravitational first principle approaches

3.5 Interesting Future Directions

In these lectures, I focused my attention in two particular corners of string theory, the half-BPS sector of $\mathcal{N} = 4$ SYM and AdS_3/CFT_2, to highlight recent developments in the field, but always attempting to describe them in the bigger context of some of the most fundamental questions in theoretical physics, especially when involving gravity issues. Such broader frame is summarised in Fig. 3.6.

There are many important foundational questions that remain open. Given the topics covered in these notes, it is natural to highlight the following subset

- Develop new ideas to understand the microscopics of non-extremal black holes and whether there exists any relation with the physics of Rindler space (its near-horizon geometry).
- Develop gauge theory techniques to compute correlations functions in the AdS/CFT correspondence in non-trivial heavy states to achieve a more accurate mathematical formulation of the information paradox [92] along the lines outlined in Sect. 3.2.1.
- Develop the formulation of holography in de Sitter and Minkowski spacetimes. Apply some of the black hole ideas reviewed here to more general holographic scenarios such as time dependent ones, cosmology and their classical singularities. It is tempting to speculate that the Big Bang singularity is a consequence of the loss of quantum information regarding the initial state of the universe. Entanglement entropy [51] and its AdS/CFT formulation [136] could provide some technical tools to handle some of these time dependent questions.

There is a large body of literature giving evidence, from different perspectives, that classical gravity is thermodynamical in nature. The universal connection between thermodynamics and statistical mechanics led the way to the notion of holography. The duality between open and closed strings led to the discovery of the AdS/CFT correspondence. They all definitely shed new light in the fascinating quest to understand what gravity and spacetime are. The future will bring, no doubt, further surprises and revelations in the resolution of some of the more fundamental puzzles in theoretical physics.

References

1. L.F. Abbott, S. Deser, Nucl. Phys. B **195**, 76 (1982).
   V. Iyer, R.M. Wald, Phys. Rev. D **50**, 846 (1994).
   I.M. Anderson, C.G. Torre, Phys. Rev. Lett. **77**, 4109 (1996).
   C.G. Torre, Local cohomology in field theory with applications to the Einstein equations. arXiv:hep-th/9706092.
   G. Barnich, F. Brandt, M. Henneaux, Commun. Math. Phys. **174**, 57 (1995).
   G. Barnich, F. Brandt, M. Henneaux, Phys. Rep. **338**, 439 (2000)
2. A.J. Amsel, G.T. Horowitz, D. Marolf, M.M. Roberts, J. High Energy Phys. **0909**, 044 (2009)
3. N. Arkani-Hamed, S. Dubovsky, A. Nicolis, E. Trincherini, G. Villadoro, J. High Energy Phys. **0705**, 055 (2007)
4. T. Azeyanagi, N. Ogawa, S. Terashima, Emergent AdS_3 in the zero entropy extremal black Holes. arXiv:1010.4291 (hep-th).
   Y. Matsuo, T. Nishioka, New near-horizon limit in Kerr/CFT. arXiv:1010.4549 (hep-th).
   T. Azeyanagi, N. Ogawa, S. Terashima, On non-chiral extension of Kerr/CFT. (arXiv:1102.3423 (hep-th))
5. H. Bacry, A. Grossmann, J. Zak, Phys. Rev. B **12**, 1118 (1975)
6. V. Balasubramanian, B. Czech, Quantitative approaches to information recovery from black holes. (arXiv:1102.3566 (hep-th))
7. V. Balasubramanian, R. Gopakumar, F. Larsen, Nucl. Phys. B **526**, 415 (1998)
8. V. Balasubramanian, P. Kraus, A.E. Lawrence, Phys. Rev. D **59**, 046003 (1999)
9. V. Balasubramanian, M. Berkooz, A. Naqvi, M.J. Strassler, J. High Energy Phys. **0204**, 034 (2002)
10. V. Balasubramanian, A. Naqvi, J. Simon, J. High Energy Phys. **0408**, 023 (2004)
11. V. Balasubramanian, V. Jejjala, J. Simon, Int. J. Mod. Phys. **D14**, 2181–2186 (2005)
12. V. Balasubramanian, P. Kraus, M. Shigemori, Class. Quantum Gravity **22**, 4803 (2005)
13. V. Balasubramanian, J. de Boer, V. Jejjala, J. Simon, J. High Energy Phys. **0512**, 006 (2005)
14. V. Balasubramanian, B. Czech, K. Larjo, J. Simon, J. High Energy Phys. **0611**, 001 (2006)
15. V. Balasubramanian, D. Marolf, M. Rozali, Gen. Relativ. Gravit. **38**, 1529 (2006). (Int. J. Mod. Phys. D **15**, 2285 (2006))
16. V. Balasubramanian, B. Czech, K. Larjo, D. Marolf, J. Simon, J. High Energy Phys. **0712**, 067 (2007)
17. V. Balasubramanian, J. de Boer, S. El-Showk, I. Messamah, Class. Quantum Gravity **25**, 214004 (2008)
18. V. Balasubramanian, B. Czech, V.E. Hubeny, K. Larjo, M. Rangamani, J. Simon, Gen. Relativ. Gravit. **40**, 1863 (2008)
19. V. Balasubramanian, J. de Boer, V. Jejjala, J. Simon, J. High Energy Phys. **0805**, 067 (2008)
20. V. Balasubramanian, J. de Boer, M. Sheikh-Jabbari, J. Simón, J. High Energy Phys. **1002**, 017 (2010)
21. V. Balasubramanian, J. Parsons, S.F. Ross, Class. Quantum Gravity **28**, 045004 (2011)
22. M. Banados, C. Teitelboim, J. Zanelli, Phys. Rev. Lett. **69**, 1849 (1992)
23. M. Banados, M. Henneaux, C. Teitelboim, J. Zanelli, Phys. Rev. D **48**, 1506 (1993)

24. T. Banks, W. Fischler, S.H. Shenker, L. Susskind, Phys. Rev. **D55**, 5112–5128 (1997)
25. J.M. Bardeen, G.T. Horowitz, Phys. Rev. D **60**, 104030 (1999)
26. J.M. Bardeen, B. Carter, S.W. Hawking, Commun. Math. Phys. **31**, 161 (1973)
27. V. Bargmann, P. Butera, L. Girardello, J.R. Klauder, Rep. Math. Phys. **2**, 221 (1971)
28. G. Barnich, Class. Quantum Gravity **20**, 3685 (2003)
29. G. Barnich, F. Brandt, Nucl. Phys. B **633**, 3 (2002)
30. G. Barnich, G. Compere, J. Math. Phys. **49**, 042901 (2008)
31. B. Bates, F. Denef, Exact solutions for supersymmetric stationary black hole composites. arXiv:hep-th/0304094
32. M. Becker, S. Cremonini, W. Schulgin, J. High Energy Phys. **1009**, 022 (2010). M. Becker, S. Cremonini, W. Schulgin, J. High Energy Phys. **1102**, 007 (2011)
33. K. Behrndt, A.H. Chamseddine, W.A. Sabra, Phys. Lett. **B442**, 97–101 (1998). K. Behrndt, M. Cvetic, W.A. Sabra, Nucl. Phys. **B553**, 317–332 (1999)
34. J.D. Bekenstein, Phys. Rev. D **7**, 2333 (1973)
35. I. Bena, N.P. Warner, Phys. Rev. D **74**, 066001 (2006)
36. I. Bena, N.P. Warner, Lect. Notes Phys. **755**, 1 (2008)
37. I. Bena, C.-W. Wang, N.P. Warner, J. High Energy Phys. **0611**, 042 (2006)
38. I. Bena, S. Giusto, C. Ruef, N.P. Warner, J. High Energy Phys. **0911**, 089 (2009)
39. I. Bena, S. Giusto, C. Ruef, N.P. Warner, J. High Energy Phys. **1003**, 047 (2010)
40. I. Bena, N. Bobev, S. Giusto, C. Ruef, N.P. Warner, J. High Energy Phys. **1103**, 022 (2011)
41. D. Berenstein, J. High Energy Phys. **0407**, 018 (2004)
42. D. Berenstein, J. High Energy Phys. **0601**, 125 (2006)
43. P. Berglund, E.G. Gimon, T.S. Levi, J. High Energy Phys. **0606**, 007 (2006)
44. L. Boltzmann, Kais. Akad. Wiss. Wien Math. Naturwiss. Classe **76**, 373–435 (1877). L. Boltzmann, Kais. Akad. Wiss. Wien Math. Naturwiss. Classe **66**, 275–370 (1872)
45. R. Bousso, J. High Energy Phys. **9907**, 004 (1999)
46. R. Bousso, Rev. Mod. Phys. **74**, 825–874 (2002)
47. D. Brecher, A. Chamblin, H.S. Reall, Nucl. Phys. B **607**, 155 (2001)
48. I. Bredberg, T. Hartman, W. Song, A. Strominger, J. High Energy Phys. **1004**, 019 (2010)
49. I. Bredberg, C. Keeler, V. Lysov, A. Strominger, Cargese lectures on the Kerr/CFT correspondence. arXiv:1103.2355 (hep-th)
50. J.D. Brown, M. Henneaux, Commun. Math. Phys. **104**, 207 (1986)
51. P. Calabrese, J.L. Cardy, J. Stat. Mech. **0406**, P06002 (2004)
52. J.L. Cardy, Nucl. Phys. B **270**, 186 (1986)
53. A. Castro, F. Larsen, J. High Energy Phys. **0912**, 037 (2009)
54. A. Castro, D. Grumiller, F. Larsen, R. McNees, J. High Energy Phys. **0811**, 052 (2008)
55. A. Castro, C. Keeler, F. Larsen, J. High Energy Phys. **1007**, 033 (2010)
56. A. Castro, A. Maloney, A. Strominger, Phys. Rev. D **82**, 024008 (2010)
57. D.D.K. Chow, M. Cvetic, H. Lu, C.N. Pope, Phys. Rev. D **79**, 084018 (2009)
58. B.D. Chowdhury, S.D. Mathur, Class. Quantum Gravity **25**, 135005 (2008). B.D. Chowdhury, S.D. Mathur, Class. Quantum Gravity **25**, 225021 (2008). B.D. Chowdhury, S.D. Mathur, Class. Quantum Gravity **26**, 035006 (2009)
59. G. Compere, Symmetries and conservation laws in Lagrangian gauge theories with applications to the mechanics of black holes and to gravity in three dimensions. arXiv:0708.3153 (hep-th)
60. G. Compere, K. Murata, T. Nishioka, J. High Energy Phys. **0905**, 077 (2009)
61. G. Compere, W. Song, A. Virmani, Microscopics of extremal Kerr from spinning M5 branes. (arXiv:1010.0685 (hep-th))
62. S. Corley, A. Jevicki, S. Ramgoolam, Adv. Theor. Math. Phys. **5**, 809–839 (2002)
63. O. Coussaert, M. Henneaux, Self-dual solutions of $2 + 1$ Einstein gravity with a negative cosmological constant. arXiv:hep-th/9407181
64. C. Crnkovic, E. Witten, Covariant description Of canonical formalism in geometrical theories, in *Three Hundred Years of Gravitation*, ed. by S.W. Hawking, W. Israel (Cambridge University Press, Cambridge/New York, 1987), pp. 676–684

65. M. Cvetic, F. Larsen, J. High Energy Phys. **0909**, 088 (2009)
66. M. Cvetic, H. Lu, C.N. Pope, Nucl. Phys. B **545**, 309 (1999)
67. G. Dall'Agata, S. Giusto, C. Ruef, J. High Energy Phys. **1102**, 074 (2011)
68. J. de Boer, S. El-Showk, I. Messamah, D.V.d. Bleeken, Quantizing N = 2 multicenter solutions. arXiv:0807.4556 (hep-th)
69. J. de Boer, S. El-Showk, I. Messamah, D.V.d. Bleeken, A bound on the entropy of supergravity? arXiv:0906.0011 (hep-th)
70. J. de Boer, M.M. Sheikh-Jabbari, J. Simon, Near horizon limits of massless BTZ and their CFT duals. arXiv:1011.1897 (hep-th)
71. L. D'Errico, W. Mueck, R. Pettorino, J. High Energy Phys. **0705**, 063 (2007)
72. R. de Mello Koch, N. Ives, M. Stephanou, Phys. Rev. **D79**, 026004 (2009)
73. R. de Mello Koch, T.K. Dey, N. Ives, M. Stephanou, J. High Energy Phys. **0908**, 083 (2009)
74. F. Denef, G.W. Moore, Split states, entropy enigmas, holes and halos. (hep-th/0702146 (HEP-TH))
75. A. Dhar, G. Mandal, N.V. Suryanarayana, J. High Energy Phys. **0601**, 118 (2006)
76. O.J.C. Dias, H.S. Reall, J.E. Santos, J. High Energy Phys. **0908**, 101 (2009)
77. R. Dijkgraaf, J.M. Maldacena, G.W. Moore, E.P. Verlinde, A black hole farey tail. arXiv: hep-th/0005003
78. A. Einstein, Ann. Phys. **49**, 769 (1916). (Ann. Phys. **14**, 517 (2005))
79. R. Fareghbal, C.N. Gowdigere, A.E. Mosaffa, M.M. Sheikh-Jabbari, J. High Energy Phys. **0808**, 070 (2008)
80. R. Fareghbal, C.N. Gowdigere, A.E. Mosaffa, M.M. Sheikh-Jabbari, arXiv:0805.0203 (hep-th)
81. J.M. Figueroa-O'Farrill, J. Simon, Adv. Theor. Math. Phys. **8**, 217 (2004)
82. V.P. Frolov, K.S. Thorne, Phys. Rev. **D39**, 2125 (1989)
83. W.D. Goldberger, J. High Energy Phys. **0903**, 069 (2009)
84. L. Grant, L. Maoz, J. Marsano, K. Papadodimas, V.S. Rychkov, J. High Energy Phys. **0508**, 025 (2005)
85. S.S. Gubser, I.R. Klebanov, A.M. Polyakov, Phys. Lett. B **428**, 105 (1998)
86. M. Guica, A. Strominger, J. High Energy Phys. **1102**, 010 (2011)
87. M. Guica, T. Hartman, W. Song, A. Strominger, Phys. Rev. D **80**, 124008 (2009)
88. R.K. Gupta, A. Sen, J. High Energy Phys. **0904**, 034 (2009)
89. T. Hartman, A. Strominger, J. High Energy Phys. **0904**, 026 (2009)
90. T. Hartman, K. Murata, T. Nishioka, A. Strominger, J. High Energy Phys. **0904**, 019 (2009)
91. S.W. Hawking, Commun. Math. Phys. **43**, 199 (1975) (Erratum-ibid. **46**, 206 (1976))
92. S.W. Hawking, Phys. Rev. **D14**, 2460–2473 (1976)
93. S.W. Hawking, G.F.R. Ellis, *The Large Scale Structure of Space-Time* (Cambridge University Press, Cambridge, 1973)
94. S.W. Hawking, R. Penrose, Proc. R. Soc. Lond. **A314**, 529–548 (1970)
95. I. Heemskerk, J. Polchinski, Holographic and Wilsonian renormalization groups. arXiv:1010.1264 (hep-th)
    T. Faulkner, H. Liu, M. Rangamani, Integrating out geometry: Holographic Wilsonian RG and the membrane paradigm. arXiv:1010.4036 (hep-th)
96. S. Hellerman, J. Polchinski, Phys. Rev. **D59**, 125002 (1999)
97. C.P. Herzog, J. Phys. A **42**, 343001 (2009).
    S.A. Hartnoll, Class. Quantum Gravity **26**, 224002 (2009).
    G.T. Horowitz, Introduction to holographic superconductors. arXiv:1002.1722 (hep-th).
    J. McGreevy, Adv. High Energy Phys. **2010**, 723105 (2010).
    T. Faulkner, H. Liu, J. McGreevy, D. Vegh, Emergent quantum criticality, Fermi surfaces, and AdS2. arXiv:0907.2694 (hep-th)
98. M. Hillery, R.F. O'Connell, M.O. Scully, E.P. Wigner, Phys. Rep. **106**(3), 121–167 (1984)
99. C.M. Hull, P.K. Townsend, Nucl. Phys. B **438**, 109 (1995)
100. V. Jejjala, S. Nampuri, J. High Energy Phys. **1002**, 088 (2010)
101. V. Jejjala, O. Madden, S.F. Ross, G. Titchener, Phys. Rev. **D71**, 124030 (2005)

102. J. Kinney, J.M. Maldacena, S. Minwalla, S. Raju, Commun. Math. Phys. **275**, 209 (2007)
103. H.K. Kunduri, J. Lucietti, H.S. Reall, Class. Quantum Gravity **24**, 4169 (2007)
104. P. Kraus, H. Ooguri, S. Shenker, Phys. Rev. D **67**, 124022 (2003).
     L. Fidkowski, V. Hubeny, M. Kleban, S. Shenker, J. High Energy Phys. **0402**, 014 (2004).
     T.S. Levi, S.F. Ross, Phys. Rev. D **68**, 044005 (2003).
     V. Balasubramanian, T.S. Levi, Phys. Rev. D **70**, 106005 (2004).
     D. Brecher, J. He, M. Rozali, J. High Energy Phys. **0504**, 004 (2005).
     G. Festuccia, H. Liu, J. High Energy Phys. **0604**, 044 (2006).
     B. Freivogel, V.E. Hubeny, A. Maloney, R. Myers, M. Rangamani, S. Shenker, J. High Energy Phys. **0603**, 007 (2006).
     K. Maeda, M. Natsuume, T. Okamura, Phys. Rev. D **74**, 046010 (2006).
     A. Hamilton, D. Kabat, G. Lifschytz, D.A. Lowe. arXiv:hep-th/0612053
105. H. Lin, O. Lunin, J. Maldacena, J. High Energy Phys. **0410**, 025 (2004)
106. F. Loran, H. Soltanpanahi, Class. Quantum Gravity **26**, 155019 (2009)
107. H. Lu, J. Mei, C. N. Pope, J. High Energy Phys. **0904**, 054 (2009).
     T. Azeyanagi, N. Ogawa, S. Terashima, J. High Energy Phys. **0904**, 061 (2009)
108. O. Lunin, S.D. Mathur, Nucl. Phys. B **623**, 342 (2002)
109. O. Lunin, S.D. Mathur, Phys. Rev. Lett. **88**, 211303 (2002)
110. O. Lunin, J.M. Maldacena, L. Maoz, Gravity solutions for the D1-D5 system with angular momentum. arXiv:hep-th/0212210
111. J.M. Maldacena, Adv. Theor. Math. Phys. **2**, 231 (1998). (Int. J. Theor. Phys. **38**, 1113 (1999))
112. J.M. Maldacena, J. High Energy Phys. **0304**, 021 (2003).
     L. Dyson, M. Kleban, L. Susskind, J. High Energy Phys. **0210**, 011 (2002).
     L. Dyson, J. Lindesay, L. Susskind, J. High Energy Phys. **0208**, 045 (2002).
     N. Goheer, M. Kleban, L. Susskind, J. High Energy Phys. **0307**, 056 (2003).
     D. Birmingham, I. Sachs, S.N. Solodukhin, Phys. Rev. D **67**, 104026 (2003).
     J.L.F. Barbon, E. Rabinovici, J. High Energy Phys. **0311**, 047 (2003).
     J.L.F. Barbon, E. Rabinovici, Fortschr. Phys. **52**, 642–649 (2004).
     J.L.F. Barbon, E. Rabinovici, Topology change and unitarity in quantum black hole dynamics. arXiv:hep-th/0503144.
     M. Kleban, M. Porrati, R. Rabadan, J. High Energy Phys. **0410**, 030 (2004)
113. J.M. Maldacena, H. Ooguri, J. Math. Phys. **42**, 2929–2960 (2001)
114. J.M. Maldacena, J. Michelson, A. Strominger, J. High Energy Phys. **9902**, 011 (1999)
115. J.M. Maldacena, H. Ooguri, J. Son, J. Math. Phys. **42**, 2961–2977 (2001).
     S. Hemming, E. Keski-Vakkuri, P. Kraus, J. High Energy Phys. **0210**, 006 (2002)
116. J. Maldacena, D. Martelli, Y. Tachikawa, J. High Energy Phys. **0810**, 072 (2008)
117. G. Mandal, J. High Energy Phys. **0508**, 052 (2005)
118. L. Maoz, V.S. Rychkov, J. High Energy Phys. **0508**, 096 (2005)
119. E.J. Martinec, W. McElgin, J. High Energy Phys. **0204**, 029 (2002)
120. S.D. Mathur, Fortschr. Phys. **53**, 793 (2005)
121. S.D. Mathur, Class. Quantum Gravity **23**, R115 (2006)
122. S.D. Mathur, Lect. Notes Phys. **769**, 3–48 (2009)
123. S. D. Mathur, Class. Quantum Gravity **26**, 224001 (2009)
124. S.D. Mathur, Fuzzballs and the information paradox: a summary and conjectures. arXiv:0810.4525 (hep-th)
125. S.D. Mathur, The information paradox and the infall problem. (arXiv:1012.2101 (hep-th))
126. S.D. Mathur, C.J. Plumberg, Correlations in Hawking radiation and the infall problem. (arXiv:1101.4899 (hep-th))
127. J. McGreevy, L. Susskind, N. Toumbas, J. High Energy Phys. **0006**, 008 (2000)
128. R.C. Myers, J. High Energy Phys. **9912**, 022 (1999)
129. R.C. Myers, O. Tafjord, J. High Energy Phys. **0111**, 009 (2001)
130. J. Navarro-Salas, P. Navarro, Nucl. Phys. **B579**, 250–266 (2000).
     M. Cadoni, S. Mignemi, Phys. Lett. **B490**, 131–135 (2000)
131. D.N. Page, Phys. Rev. Lett. **71**, 3743 (1993)

132. R. Penrose, Phys. Rev. Lett. **14**, 57–59 (1965)
133. A.M. Perelomov, Teor. Mat. Fiz. **6**, 213 (1971)
134. J. Polchinski, Phys. Rev. Lett. **75**, 4724 (1995)
135. V.S. Rychkov, J. High Energy Phys. **0601**, 063 (2006)
136. S. Ryu, T. Takayanagi, Phys. Rev. Lett. **96**, 181602 (2006).
     S. Ryu, T. Takayanagi, J. High Energy Phys. **0608**, 045 (2006).
     S. Ryu, T. Takayanagi, J. Phys. A **A42**, 504008(2009)
137. N. Seiberg, Phys. Rev. Lett. **79**, 3577 (1997)
138. A. Sen, J. High Energy Phys. **0509**, 038 (2005)
139. A. Sen, Gen. Relativ. Gravit. **40**, 2249 (2008)
140. A. Sen, J. High Energy Phys. **0811**, 075 (2008)
141. A. Sen, Int. J. Mod. Phys. A **24**, 4225 (2009)
142. A. Sen, J. High Energy Phys. **0908**, 068 (2009)
143. A. Sen, J. High Energy Phys. **1005**, 097 (2010)
144. C.E. Shannon, Bell Syst. Tech. J. **27**, 379–423 (1948)
145. J. Simon, Phys. Rev. D **81**, 024003 (2010)
146. K. Skenderis, M. Taylor, Phys. Rep. **467**, 117 (2008)
147. A. Strominger, J. High Energy Phys. **9802**, 009 (1998)
148. A. Strominger, J. High Energy Phys. **9901**, 007 (1999)
149. A. Strominger, C. Vafa, Phys. Lett. B **379**, 99 (1996)
150. L. Susskind, J. Math. Phys. **36**, 6377 (1995)
151. Y. Takayama, A. Tsuchiya, J. High Energy Phys. **0510**, 004 (2005)
152. G. 't Hooft, Dimensional reduction in quantum gravity. arXiv:gr-qc/9310026
153. J.A. Wheeler, Phys. Rev. **97**, 511 (1955).
     J.A. Wheeler, Ann. Phys. **2**, 604 (1957)
154. E. Witten, Nucl. Phys. B **443**, 85 (1995)
155. E. Witten, Adv. Theor. Math. Phys. **2**, 253 (1998)

Chapter 4
Issues on Black Holes in Four Dimensional Supergravity

L. Andrianopoli, R. D'Auria and M. Trigiante

4.1 Introduction

Black holes have played a major role in understanding non-perturbative aspects of string/M-theory [5]. Reproducing the Bekenstein-Hawking "area law" formula for the black hole entropy through direct microstate counting for specific static BPS black holes, has been one of the major successes of string theory [36]. This was largely possible due to the residual supersymmetry of the solution, which made the degeneracy of its microstates independent of the scalar moduli coupled to it, including the string coupling constant. In this way the result of the string microstate counting, only possible in the weak coupling regime, could be compared with the area-law prediction for the entropy in the supergravity limit (large charges). Extremal (i.e. zero-temperature) static regular black holes include the BPS solutions as a subclass and share with them some common properties. Indeed they all exhibit the *attractor mechanism* [21]: The scalar fields which are coupled to the solution flow from their values at radial infinity towards a fixed point in the scalar manifold, in correspondence to the horizon of the solution, which only depends on the quantized charges. The near horizon geometry of four dimensional regular extremal solutions has the form $AdS_2 \times S_2$, where the S_2 factor describes the horizon of the solution and its radius R_H only depends on the quantized charges. The independence of the near horizon geometry on the moduli at infinity, and thus on the string coupling constant, suggests that the statistical entropy computed by string microstate counting in the weak string-coupling regime, can reproduce the expression for the entropy evaluated on extremal supergravity solutions. The extremality property of a solution seems to be, in this respect, more fundamental than supersymmetry [18].

L. Andrianopoli · R. D'Auria · M. Trigiante (✉)
Laboratory of Theoretical Physics, Dipartimento di Scienze Applicate e Tecnologia,
Politecnico di Torino, C.so Duca degli Abruzzi, 10129 Torino, Italy
e-mail: laura.andrianopoli@polito.it; riccardo.dauria@polito.it; mario.trigiante@polito.it

S. Bellucci (ed.), *Supersymmetric Gravity and Black Holes*, Springer Proceedings   143
in Physics 142, DOI 10.1007/978-3-642-31380-6_4,
© Springer-Verlag Berlin Heidelberg 2013

Dualities have played a major role in understanding the non-perturbative aspects of superstring theory [38]. They define (non-perturbative) relations between various superstring-theories realized on different backgrounds, as if they were distinct descriptions of the same microscopic degrees of freedom. The known dualities were conjectured [31] to be encoded in the global symmetry group (*duality group*) of the supergravity model describing the low energy dynamics of string/M-theory compactifications. It is natural to expect the entropy of a black hole, which counts the number of micro-states realizing the given macroscopic state (i.e. the supergravity solution), to be independent of the description of the former, namely to be described by a duality-invariant quantity. This is consistent with the property of the entropy of extremal static, asymptotically flat black holes in supergravity, computed in terms of the area of their horizon, to be described by a function of the quantized electric and magnetic charges, which is invariant with respect to the duality group of the classical theory (see [5] and references therein).

In this report we shall restrict ourselves to extremal, spherically symmetric and asymptotically flat black holes in four dimensional extended supergravities. It is possible to associate with these solutions a characteristic function W of the scalar fields of the theory and of the quantized electric-magnetic charges, known in the literature as *fake-superpotential* [4, 6, 15, 16], which monotonically interpolates between the ADM mass at radial infinity and the entropy at the horizon (minimum value). This function allows to describe the radial evolution of the scalar fields and the warp factor of the metric by means of a first-order *gradient-flow* system of equations, which exhibits an asymptotically stable equilibrium point (*stable attractor*) in correspondence to the horizon of the solution, where the function W has a minimum. Being W invariant, just as the entropy is, under the action of the global symmetries of the classical model (duality invariant) [4, 6, 16], it is natural to expect it to be related to some intrinsic microscopic properties of the solution. For the same reason one would in general expect all the duality invariant properties of extremal solutions to be of special physical relevance. It is useful in this respect to classify extremal solutions in orbits of the duality group of the supergravity model, each labeled by different values of duality invariants [10]. Each duality orbit can be conveniently characterized in terms of a *generating* (or *seed*) solution, defined as the simplest solution which captures all the duality invariant features of the most general one (entropy, flat directions etc.).[1]

Here we wish to review some general properties of the W function which naturally arise from its identification with the Hamilton's characteristic function of a suitable autonomous Hamiltonian system [6].

This report is organized as follows. In Sect. 4.2 we review the basic facts about the global on-shell symmetries of extended supergravities. In Sect. 4.3 we focus on spherically symmetric, asymptotically flat black hole solutions and introduce the autonomous Hamiltonian system describing the radial flow of the warp function

---

[1] The generating solutions for BPS and extremal non-BPS black holes were constructed in [3] and [27], respectively.

and of the scalar fields coupled to the solution. This will allow us to review the characterization of the *fake-superpotential* as a solution to the Hamilton-Jacobi equation and to illustrate some of its general features such as duality-invariance and flat-directions. We shall then comment on the duality orbit with $I_4 < 0$ for which a closed-form expression for W is not known yet. In Sect. 4.5 we review the basic facts about the description of spherically symmetric $D = 4$ black holes in terms of geodesics on the scalar manifold of a suitable Euclidean $D = 3$ theory. We shall define the W-function in this new setting and relate it to the superpotential in the $D = 4$ theory. We end with few concluding remarks.

4.2 Duality in Extended Supergravities

We shall restrict ourselves to four-dimensional extended (i.e. $N \geq 2$) supergravities. The bosonic section consists of the graviton $g_{\mu\nu}(x)$, n_V vector fields $A_\mu^\Lambda(x)$, $\Lambda = 0, \ldots, n_V - 1$, and n_s scalar fields $\phi^r(x)$, $r = 1, \ldots, n_s$. The latter are described by a non-linear sigma-model, namely are coordinates of a non-compact, Riemannian, simply-connected manifold $\mathcal{M}_{scal}$. For $N > 2$ this manifold is homogeneous symmetric of the form $\mathcal{M}_{scal} = G/H$, where G is the semisimple isometry Lie group and H its maximal compact subgroup. In the $N = 2$ case (eight supercharges) $\mathcal{M}_{scal}$ need not even be homogeneous. We shall however restrict ourselves to models exhibiting a homogeneous symmetric scalar manifold of the form $\mathcal{M}_{scal} = G/H$ (symmetric models). The action of an isometry transformation $g \in G$ on the scalar fields ϕ^r parametrizing $\mathcal{M}_{scal}$ is defined by means of a *coset representative* $\mathbb{L}(\phi) \in G/H$ as follows:

$$g \cdot \mathbb{L}(\phi^r) = \mathbb{L}(g \star \phi^r) \cdot h(\phi^r, g), \tag{4.1}$$

where $g \star \phi^r$ denote the transformed scalar fields, non-linear functions of the original ones ϕ^r, and $h(\phi^r, g)$ is a *compensator* in H. The coset representative is defined modulo right action of H and is fixed by the chosen parametrization of the manifold.

We shall also limit our analysis to *ungauged* supergravities, namely to models in which the vector fields are not minimally coupled to any other field. All these models have a $U(1)^{n_V}$ gauge symmetry with respect to which all fields are neutral and the vector fields transform as gauge connections: $A_\mu^\Lambda \to A_\mu^\Lambda + \partial_\mu \xi^\Lambda$.

Setting $\kappa^2 = 8\pi G_N = \hbar = c' = 1$, ($c'$ being the speed of light, not to be confused with the *extremality parameter* to be introduced in the following), the bosonic action reads:

$$S_4 = \int d^4x \sqrt{|g|} \, \mathcal{L}_4 = \int d^4x \sqrt{|g|} \left[\frac{R[g]}{2} - \frac{1}{2} G_{rs}(\phi) \, \partial_\mu \phi^r \partial^\mu \phi^s \right.$$

$$\left. + \frac{1}{4} F_{\mu\nu}^\Lambda I_{\Lambda\Sigma}(\phi) \, F^{\Sigma\,\mu\nu} + \frac{1}{8\sqrt{|g|}} \, \epsilon_{\mu\nu\rho\sigma} F^{\Lambda\,\mu\nu} R_{\Lambda\Sigma}(\phi) \, F^{\Sigma\,\rho\sigma} \right], \tag{4.2}$$

where $|g| \equiv |\det(g_{\mu\nu})|$ and we are using the "mostly plus" signature for the metric $g_{\mu\nu}$.[2] The scalar-dependent matrix $G_{rs}(\phi)$ represents the positive definite metric on $\mathcal{M}_{scal}$, where we have collectively denoted the scalar fields by the short-hand notation $\phi \equiv (\phi^r)$. The vector field strengths are defined as usual: $F^{\Lambda}_{\mu\nu} \equiv \partial_{\mu} A^{\Lambda}_{\nu} - \partial_{\nu} A^{\Lambda}_{\mu}$. The above action shows a general feature of all supergravity theories: The scalar fields are non-minimally coupled to the vector fields through the symmetric $n_V \times n_V$ matrices $I_{\Lambda\Sigma}(\phi)$, $R_{\Lambda\Sigma}(\phi)$. The former enters the kinetic term of the vector fields as a *generalized coupling constant matrix*, while the latter defines a generalized *theta-term*. The absence of ghosts requires the coupling constant matrix $I_{\Lambda\Sigma}(\phi)$ to be negative definite.

The scalar fields are defined by the *solvable parametrization* of the manifold, that is the isotropy group H can be fixed at each point of the manifold in such a way that the coset representative $\mathbb{L}(\phi)$ belong to the solvable Lie group of the form $\exp(Solv_4)$, defined by the Iwasawa decomposition of G with respect to H. The scalar fields are then parameters of the solvable Lie algebra $Solv_4$:

$$\mathbb{L}(\phi^r) = e^{\phi^r T_r} \in \exp(Solv_4) , \qquad (4.3)$$

where $\{T_r\}_{r=1,\dots,n_s}$ is a basis of $Solv_4$. On $Solv_4$ a positive definite scalar product $(\cdot,\cdot) : Solv_4 \times Solv_4 \to \mathbb{R}$ can be defined which induces a metric on the group manifold $\exp(Solv_4)$ that reproduces the Riemannian metric on $\mathcal{M}_{scal}$:

$$ds^2 = d\phi^r G_{rs}(\phi) d\phi^s = \left(\mathbb{L}^{-1}d\mathbb{L}(\phi), \mathbb{L}^{-1}d\mathbb{L}(\phi)\right) . \qquad (4.4)$$

The above description also applies to homogeneous non–symmetric scalar manifolds in $N = 2$ which however admit a solvable isometry group with a simple, transitive action on it [1]. Note that, having defined the scalar fields as the parameters of $\exp(Solv_4)$, the solvable parametrization is required to be global on the scalar manifold $\mathcal{M}_{scal}$. This is possible only if $\mathcal{M}_{scal}$ is simply connected since the manifold $\exp(Solv_4)$ does not have unshrinkable cycles.[3]

It is useful to introduce the dual field strengths $G_{\Lambda\,\mu\nu}$ defined as:

$$G_{\Lambda\,\mu\nu} \equiv -\sqrt{|g|}\,\epsilon_{\mu\nu\rho\sigma}\frac{\partial \mathcal{L}_4}{\partial F^{\Lambda}_{\rho\sigma}} = R_{\Lambda\Sigma}\,F^{\Sigma}_{\mu\nu} - I_{\Lambda\Sigma}\,{}^*F^{\Sigma}_{\mu\nu} , \qquad (4.5)$$

where

$$ {}^*F^{\Sigma}_{\mu\nu} \equiv \frac{\sqrt{|g|}}{2}\,\epsilon_{\mu\nu\rho\sigma}\,F^{\Lambda\,\rho\sigma} . \qquad (4.6)$$

---

[2] We also use the convention $\epsilon_{0123} = 1$.

[3] If the solvable group manifold contained a non-trivial cycle, this would be generated by a compact isometry which is absent in $Solv_4$.

The equations of motion for the scalar and vector fields read:

$$D_\mu(\partial^\mu \phi^r) = \frac{1}{4} G^{rs} \left[F^\Lambda_{\mu\nu} \partial_s I_{\Lambda\Sigma} F^{\Sigma\,\mu\nu} + F^\Lambda_{\mu\nu} \partial_s R_{\Lambda\Sigma} {}^* F^{\Sigma\,\mu\nu} \right],$$

$$\nabla_\mu({}^* F^{\Lambda\,\mu\nu}) = 0 ; \quad \nabla_\mu({}^* G^{\Lambda\,\mu\nu}) = 0, \tag{4.7}$$

where $\partial_s \equiv \frac{\partial}{\partial \phi^s}$, ∇_μ is the covariant derivative containing the Levi-Civita connection on space-time, while D_μ also contains the connection Ω on $\mathcal{M}_{scal}$:

$$D_\mu \equiv \nabla_\mu + \partial_\mu \phi^r \, \Omega_r . \tag{4.8}$$

Using (4.5) and the property that $^{**}F^\Lambda = -F^\Lambda$, we can express $^* F^\Lambda$ and $^* G_\Lambda$ as linear functions of F^Λ and G_Λ:

$$^* F^\Lambda = I^{-1\,\Lambda\Sigma} (R_{\Sigma\Gamma} F^\Gamma - G_\Sigma) ; \quad {}^* G_\Lambda = (RI^{-1} R + I)_{\Lambda\Sigma} F^\Sigma - (RI^{-1})^\Sigma_\Lambda G_\Sigma, \tag{4.9}$$

where, for the sake of simplicity, we have omitted the space-time indices. It is useful to arrange F^Λ and G_Λ in a single $2n_V$-dimensional vector $\mathbb{F} \equiv (\mathbb{F}^M)$ of two-forms:

$$\mathbb{F}_{\mu\nu} \equiv \begin{pmatrix} F^\Lambda_{\mu\nu} \\ G_{\Lambda\mu\nu} \end{pmatrix}, \tag{4.10}$$

in terms of which Eq. (4.9) are easily rewritten in the following compact form:

$$^* \mathbb{F} = -\mathbb{C}\mathcal{M}(\phi^r)\,\mathbb{F}, \tag{4.11}$$

where

$$\mathbb{C} = (\mathbb{C}^{MN}) \equiv \begin{pmatrix} \mathbf{0} & \mathbf{1} \\ -\mathbf{1} & \mathbf{0} \end{pmatrix}, \tag{4.12}$$

$\mathbf{1}, \mathbf{0}$ being the $n_V \times n_V$ identity and zero-matrices, respectively, and

$$\mathcal{M}(\phi) = (\mathcal{M}(\phi)_{MN}) \equiv \begin{pmatrix} (RI^{-1} R + I)_{\Lambda\Sigma} & -(RI^{-1})^\Gamma_\Lambda \\ -(I^{-1} R)^\Delta_\Sigma & I^{-1\,\Delta\Gamma} \end{pmatrix}, \tag{4.13}$$

is a symmetric, negative-definite matrix, function of the scalar fields.

While the Maxwell equations $\nabla_\mu(^* \mathbb{F}^{M\,\mu\nu}) = 0$ are invariant with respect to a generic linear transformation on $\mathbb{F}$, the definition of G_Λ or, equivalently, Eq. (4.11) is not. On the other hand the isometry group G is a global symmetry of the scalar action, but it will in general alter the action for the vector fields as a consequence of the scalar field-dependence of the matrices I and R.

One of the most intriguing features of extended supergravities is the fact that the global invariance of the scalar action, described by G, can be extended to a global symmetry of the full set of equations of motion and Bianchi identities [24] (though not in general of the whole action). This is possible by virtue of the existence of a *flat* $Sp(2n_V, \mathbb{R})$-*bundle* on $\mathcal{M}_{scal}$ which allows to associate with each isometry $g \in G$ a symplectic matrix $\mathcal{D}[g] = (\mathcal{D}[g]_N^M)$ acting on the fiber. In other words there exists an embedding $\mathcal{D}$ of G inside $Sp(2n_V, \mathbb{R})$

$$G \xrightarrow{\mathcal{D}} Sp(2n_V, \mathbb{R}) \quad \Leftrightarrow \quad g \in G \rightarrow \mathcal{D}[g] \in Sp(2n_V, \mathbb{R}),$$

$$\mathcal{D}[g]_N^M \mathbb{C}^{NP} \mathcal{D}[g]_P^L = \mathbb{C}^{ML} \quad \Leftrightarrow \quad \mathcal{D}[g]_N^M \mathbb{C}_{ML} \mathcal{D}[g]_P^L = \mathbb{C}_{NP}, \quad (4.14)$$

where in the second line we have written the general property defining a symplectic matrix: $\mathcal{D}[g]\mathbb{C}\mathcal{D}[g]^T = \mathcal{D}[g]^T \mathbb{C}\mathcal{D}[g] = \mathbb{C}$, $(\mathbb{C}_{MN})$ having the same matrix form as the matrix $(\mathbb{C}^{MN})$ in (4.12).

The symplectic bundle completely fixes the non-minimal coupling of the scalar fields to the vectors, namely the matrices $I(\phi)$ and $R(\phi)$ or, equivalently, the matrix $\mathcal{M}(\phi)$. In particular $\mathcal{M}(\phi)$ is a *symplectic, symmetric matrix*, that is it satisfies the relation:

$$\mathcal{M}(\phi)_{MP} \mathbb{C}^{PL} \mathcal{M}(\phi)_{LN} = \mathbb{C}_{MN} \quad \Leftrightarrow \quad \mathcal{M}(\phi)^{-1} = -\mathbb{C}\mathcal{M}(\phi)\mathbb{C}, \quad (4.15)$$

and it transforms under an isometry $g \in G$ as follows:

$$\mathcal{M}(g \star \phi) = \mathcal{D}[g]^{-T} \mathcal{M}(\phi)\mathcal{D}[g]^{-1}, \quad (4.16)$$

where $\mathcal{D}[g]^{-T} \equiv (\mathcal{D}[g]^{-1})^T$. We emphasize here that the existence of a flat symplectic bundle on the scalar manifold is a general feature of all extended supergravites, including those $N = 2$ models in which the scalar manifold is not even homogeneous (i.e. the isometry group, if it exists, does not act transitively on the manifold itself). In the $N = 2$ case only the scalar fields belonging to the vector multiplets are non-minimally coupled to the vector fields, namely enter the matrices $I(\phi)$, $R(\phi)$, and they span a *special Kähler* manifold. On this manifold a flat symplectic bundle is defined,[4] which fixes the scalar dependence of the matrices $I(\phi)$, $R(\phi)$ and the matrix $\mathcal{M}(\phi)$ defined in (4.13), satisfies the properties (4.15) and (4.16).

For homogeneous manifolds we can express $\mathcal{M}(\phi)$ in terms of the coset representative:

$$\mathcal{M}(\phi)_{MN} = \mathbb{C}_{MP} \mathbb{L}(\phi)_L^P \mathbb{L}(\phi)_L^R \mathbb{C}_{RN} \quad \Leftrightarrow \quad \mathcal{M}(\phi) = \mathbb{C}\mathcal{D}[\mathbb{L}(\phi)] \, \mathcal{D}[\mathbb{L}(\phi)]^T \, \mathbb{C}, \quad (4.17)$$

---

[4] A special Kähler manifold is in general characterized by the product of a U(1)-bundle, associated with its Kähler structure (with respect to which the manifold is Hodge Kähler), and a flat symplectic bundle. See for instance [2] for an in depth account of this issue.

where summation over the index L is understood and $\mathbb{L}_L^P$ are the entries of the symplectic matrix $\mathcal{D}[\mathbb{L}(\phi)]$ associated with $\mathbb{L}(\phi)$ as an element of G. Since $\mathcal{D}$ is a homomorphism, Eq. (4.1) can also be written in terms of symplectic matrices as follows:

$$\mathcal{D}[g]\,\mathcal{D}[\mathbb{L}(\phi)] = \mathcal{D}[\mathbb{L}(g \star \phi)]\,\mathcal{D}[h(g, \phi)]\,. \tag{4.18}$$

We see that from (4.17) and (4.18), properties (4.15) and (4.16) easily follow. Let us derive (4.16):

$$\begin{aligned}
\mathcal{M}(g \star \phi) &= \mathbb{C}\mathcal{D}[\mathbb{L}(g \star \phi)]\,\mathcal{D}[\mathbb{L}(g \star \phi)]^T\,\mathbb{C} \\
&= \mathbb{C}\mathcal{D}[g]\,\mathcal{D}[\mathbb{L}(\phi)]\mathcal{D}[h]^{-1}\,\mathcal{D}[h]^{-T}\,\mathcal{D}[\mathbb{L}(\phi)]^T\,\mathcal{D}[g]^T\,\mathbb{C} \\
&= \mathcal{D}[g]^{-T}\,\mathbb{C}\mathcal{D}[\mathbb{L}(\phi)]\,\mathcal{D}[\mathbb{L}(\phi)]^T\,\mathbb{C}\,\mathcal{D}[g]^{-1}\,, \tag{4.19}
\end{aligned}$$

where we have used the property that $\mathcal{D}[g]$ is symplectic, $\mathbb{C}\mathcal{D}[g] = \mathcal{D}[g]^{-T}\,\mathbb{C}$, and that $\mathcal{D}[h] \equiv \mathcal{D}[h(g, \phi)]$ is orthogonal, being in a real representation of $U(n_V)$: $\mathcal{D}[h]^T = \mathcal{D}[h]^{-1}$. The latter property in particular implies that $\mathcal{M}(\phi)$, as defined in (4.17), is H-invariant, namely it does not depend on the choice of the coset representative, but only on the point ϕ of the manifold, as it should be.

We can now easily verify that the simultaneous action of G on the scalar fields and on the field strength vector $\mathbb{F}_{\mu\nu}^M$:

$$g \in G : \quad \begin{cases} \phi^r \to g \star \phi^r \\ \mathbb{F}_{\mu\nu}^M \to \mathbb{F}_{\mu\nu}^{\prime M} = \mathcal{D}[g]_N^M\,\mathbb{F}_{\mu\nu}^N \end{cases}, \tag{4.20}$$

is a symmetry of the scalar and vector field equations. It is clearly a symmetry of the former, being G the isometry group of the scalar manifold. We just need to show that the above transformation leaves (4.11) invariant, namely that is holds in the transformed fields as well. Using (4.20), Eq. (4.11) can indeed be written in the new quantities as follows:

$$\mathcal{D}[g]^{-1*}\mathbb{F}' = -\mathbb{C}\mathcal{D}[g]^T\mathcal{M}(g\star\phi)\mathcal{D}[g]\,\mathcal{D}[g]^{-1}\mathbb{F}' = -\mathcal{D}[g]^{-1}\mathbb{C}\mathcal{M}(g\star\phi)\mathbb{F}'\,, \tag{4.21}$$

which is equivalent to ${}^*\mathbb{F}' = -\mathbb{C}\mathcal{M}(g \star \phi)\mathbb{F}'$.

The action of G on the field strengths and their magnetic duals, defined by the symplectic embedding $\mathcal{D}$, is a *generalized electric-magnetic duality transformation*, which promotes the isometry group of the scalar manifold to a global symmetry group of the full set of field equations and Bianchi identities. For this reason G is also referred to as the *duality group* of the classical theory.[5] Note however that G will contain transformations g whose duality action $\mathcal{D}[g]$ is *non-perturbative*,

---

namely under which $F^\Lambda \to F'^\Lambda = A^\Lambda_\Sigma F^\Sigma + B^{\Lambda\Sigma} G_\Sigma$ and $G_\Lambda \to G'_\Lambda = C_{\Lambda\Sigma} F^\Sigma + D^\Sigma_\Lambda G_\Sigma$, with $C_{\Lambda\Sigma}$, $B^{\Lambda\Sigma} \neq 0$. These are not a symmetry of the action but only of the field equations and Bianchi identities (on-shell symmetry).

The relevance of the (quantum) duality group resides in the existence of important evidence that it (or a suitable extension of it) might encode all the known string/M-theory dualities [31].

Let us end this section by giving the Einstein equations in a manifestly duality invariant form:

$$\tilde{R}_{\mu\nu} - \frac{1}{2} g_{\mu\nu} R = T^{(S)}_{\mu\nu} + T^{(V)}_{\mu\nu}, \tag{4.22}$$

where the energy-momentum tensors for the scalar and vector fields can be cast in the following general form

$$T^{(S)}_{\mu\nu} = G_{rs}(\phi)\, \partial_\mu \phi^r \partial_\nu \phi^s - \frac{1}{2} g_{\mu\nu}\, G_{rs}(\phi)\, \partial_\rho \phi^r \partial^\rho \phi^s, \tag{4.23}$$

$$T^{(V)}_{\mu\nu} = -\frac{1}{2} \mathbb{F}^T_{\mu\rho} \mathcal{M}(\phi)\, \mathbb{F}^\rho_\nu. \tag{4.24}$$

The duality invariance of the space-time metric and the scalar action imply the same property for the Einstein tensor and $T^{(S)}_{\mu\nu}$. As for $T^{(V)}_{\mu\nu}$ it is manifestly duality invariant since:

$$\mathbb{F}^T_{\mu\rho} \mathcal{M}(\phi)\, \mathbb{F}^\rho_\nu = \mathbb{F}'^T_{\mu\rho} \mathcal{D}[g]^{-T}\, \mathcal{D}[g]^T \mathcal{M}(g \star \phi) \mathcal{D}[g]\, \mathcal{D}[g]^{-1} \mathbb{F}'^\rho_\nu = \mathbb{F}'^T_{\mu\rho} \mathcal{M}(g \star \phi) \mathbb{F}'^\rho_\nu. \tag{4.25}$$

4.3 Spherically Symmetric, Asymptotically Flat Black Hole Solutions

We shall now restrict our discussion to static, spherically symmetric and asymptotically flat black hole solutions. The general ansatz for the metric has the following form:

$$ds^2 = -e^{2U} dt^2 + e^{-2U} \left(\frac{c^4}{\sinh^4(c\tau)}\, d\tau^2 + \frac{c^2}{\sinh^2(c\tau)}\, d\Omega^2 \right), \tag{4.26}$$

where $U = U(\tau)$ and the coordinate τ is related to the radial coordinate r by the following relation:

$$\frac{c^2}{\sinh^2(c\tau)} = (r - r_0)^2 - c^2 = (r - r^-)(r - r^+). \tag{4.27}$$

Here $c^2 \equiv 2ST$ is the extremality parameter of the solution, with S the entropy and T the temperature of the black hole. When c is non-vanishing the black hole has two horizons located at $r^\pm = r_0 \pm c$. The outer horizon is located at $r_H = r^+$

corresponding to $\tau \rightarrow -\infty$. The extremality limit at which the two horizons coincide, $r_H = r^+ = r^- = r_0$, is $c \rightarrow 0$. Spherical symmetry further requires the scalar fields in the solution to depend only on τ: $\phi^r = \phi^r(\tau)$. The solution is also characterized by a set of electric and magnetic charges defined as follows:

$$m^\Lambda = \frac{1}{4\pi} \int_{S^2} F^\Lambda \qquad e_\Lambda = \frac{1}{4\pi} \int_{S^2} G_\Lambda, \qquad (4.28)$$

where S^2 is a spatial two-sphere in the space-time geometry of the dyonic solution (for instance, in Minkowski space-time the two-sphere at radial infinity S^2_∞). In terms of these charges the general ansatz for the electric-magnetic field strength vector $\mathbb{F}^M$ reads[6]:

$$\mathbb{F} = \begin{pmatrix} F^\Lambda_{\mu\nu} \\ G_{\Lambda\,\mu\nu} \end{pmatrix} \frac{dx^\mu \wedge dx^\nu}{2} = e^{2U} \mathbb{C} \cdot \mathcal{M}_4(\phi^r) \cdot \Gamma \, dt \wedge d\tau + \Gamma \, \sin(\theta)\, d\theta \wedge d\varphi,$$

$$\Gamma = (\Gamma^M) = \begin{pmatrix} m^\Lambda \\ e_\Lambda \end{pmatrix} = \frac{1}{4\pi} \int_{S^2} \mathbb{F}. \qquad (4.29)$$

It is straightforward to verify that the above ansatz solves the Maxwell equations. Note that the radial evolution of the scalar fields in the presence of a black hole is a feature of this kind of solutions in supergravity and is due to the non-minimal coupling of the scalars to the vector fields through the matrices $I_{\Lambda\Sigma}(\phi)$, $R_{\Lambda\Sigma}(\phi)$. We can associate with a black hole a set of scalar-dependent charges, which are the *physical charges* measured at radial infinity. They comprise the *central charges* of the supersymmetry algebra Z_{AB}, $A, B = 1, \ldots, N$, and charges Z_I associated with the vector multiplets (matter charges). These can be grouped into a complex vector related to the vector Γ of quantized charges as follows:

$$Z_{\hat{M}}(\phi^r, \Gamma) = \begin{pmatrix} Z_{AB} \\ Z_I \\ \bar{Z}^{AB} \\ \bar{Z}^I \end{pmatrix} = \mathbb{L}(\phi^r)_{\hat{M}}^N \, \mathbb{C}_{NL} \, \Gamma^L, \qquad (4.30)$$

where the matrix $(\mathbb{L}(\phi^r)_{\hat{M}}^N)$ is obtained from the real symplectic matrix $(\mathbb{L}(\phi^r)_M^N)$ by the action of a Cayley transformation on the right index:

$$\mathbb{L}(\phi^r)_{\hat{M}}^N \equiv \mathbb{L}(\phi^r)_M^N \, (\mathcal{A}^\dagger)_{\hat{M}}^M \quad \text{where} \quad \mathcal{A} \equiv \frac{1}{\sqrt{2}} \begin{pmatrix} \mathbf{1} & i\,\mathbf{1} \\ \mathbf{1} & -i\,\mathbf{1} \end{pmatrix}. \qquad (4.31)$$

---

[6]The reader can easily verify that this ansatz satisfies Eq. (4.11) by using the property $*(dt \wedge d\tau) = -e^{-2U} \sin(\theta)\, d\theta \wedge d\varphi$.

By substituting the general ansatz in the equations of motion, the resulting equations for the metric and the scalar fields, written in terms of the evolution parameter τ, take the following simple form:

$$\frac{d^2 U}{d\tau^2} = V(\phi; e, m) \, e^{2U} \,, \tag{4.32}$$

$$\frac{D^2 \phi^r}{D\tau^2} = G^{rs}(\phi) \, \frac{\partial V(\phi; e, m)}{\partial \phi^s} \, e^{2U} \,, \tag{4.33}$$

with the constraint

$$\left(\frac{dU}{d\tau} \right)^2 + \frac{1}{2} G_{rs}(\phi) \frac{d\phi^r}{d\tau} \frac{d\phi^s}{d\tau} - V(\phi; e, m) \, e^{2U} = c^2 \,, \tag{4.34}$$

where $V(\phi; \Gamma) = V(\phi; e, m)$ is a positive definite function of the scalars and of the electric and magnetic charges of the theory, defined by:

$$V(\phi; \Gamma) = -\frac{1}{2} \Gamma^T \mathcal{M}(\phi) \, \Gamma > 0 \,. \tag{4.35}$$

The second order derivative on the left hand side of the scalar field equation is defined as follows:

$$\frac{D^2 \phi^r}{D\tau^2} \equiv \ddot{\phi}^r + \Gamma^r_{st} \, \dot{\phi}^s \dot{\phi}^t \,, \tag{4.36}$$

Γ^r_{st} being the Levi-Civita connection on the scalar manifold and we have used the short-hand notation $\dot{\phi}^s \equiv \frac{d\phi^s}{d\tau}$.

The solution is then characterized by the following conditions at radial infinity $\tau = 0$:

$$\phi^r(\tau = 0) = \phi_0^r \,, \quad U(\tau = 0) = 0 \,, \tag{4.37}$$

the latter corresponding to the requirement that the asymptotic geometry be Minkowski space-time. The ADM mass of the solution is computed at radial infinity and depends on the boundary values of the scalar fields and on the electric-magnetic charges:

$$M_{ADM}(\phi_0; \Gamma) = \lim_{\tau \to 0^-} \dot{U} \,. \tag{4.38}$$

The horizon area A_H has the following general form:

$$A_H = \lim_{\tau \to -\infty} \int_{S_2} \sqrt{g_{\theta\theta} g_{\varphi\varphi}} \, d\theta \, d\varphi = \lim_{\tau \to -\infty} 4\pi \, e^{-2U} \, \frac{c^2}{\sinh^2(c\tau)} \,. \tag{4.39}$$

Regular solutions have their singularity hidden by a finite horizon: $A_H > 0$. This implies for the warp factor the following near-horizon behavior: $e^{-2U} \sim \frac{A_H}{4\pi} \frac{e^{-2c\tau}}{4c^2}$. In the extremal limit $c \to 0$ Eq. (4.27) simplifies to $\tau = -1/r$. Regularity

then requires the warp factor to exhibit the following behavior near the horizon
($\tau \to -\infty$):

$$e^{-2U} \sim \frac{A_H}{4\pi} \tau^2. \tag{4.40}$$

Regularity of the extremal solution further demands the scalar fields not to blow-up at the horizon but to flow in the limit $\tau \to -\infty$ towards finite values: $\lim_{\tau \to -\infty} \phi^r(\tau) = \phi^r_*$. General regularity conditions of the scalar fields near the horizon then imply[7]

$$\lim_{\tau \to -\infty} \tau \dot{\phi}^r(\tau) = \lim_{\tau \to -\infty} \tau^2 \ddot{\phi}^r(\tau) = 0 \Leftrightarrow \lim_{\tau \to -\infty} e^{-U} \dot{\phi}^r(\tau) = \lim_{\tau \to -\infty} e^{-2U} \ddot{\phi}^r(\tau) = 0. \tag{4.41}$$

Multiplying both sides of Eq. (4.33) times e^{-2U} and taking the near horizon limit, properties (4.41) imply that $\phi_* \equiv (\phi^r_*)$ *is an extremum of the potential* V:

$$\left. \frac{\partial V}{\partial \phi^r} \right|_{\phi_*} = 0. \tag{4.42}$$

Since in general V may not depend on all the scalar fields, but have *flat directions*, the above equations will fix those scalars along the non-flat directions as functions of the electric and magnetic charges only. As a consequence, the value of V at the extremum will just depend on the electric and magnetic charges: $V_{ex} = V(\phi_*; e, m) = V_{ex}(e, m)$. From Eqs. (4.32) and (4.40) it follows that the area of the horizon can be expressed trough $V_{ex}(e, m)$ in terms of the electric and magnetic charges only:

$$A_H = 4\pi V_{ex}(e, m) = A_H(e, m). \tag{4.43}$$

The near horizon geometry can be easily computed, using (4.40), to have the form $AdS_2 \times S_2$, the metric only depending on the area A_H of the horizon S_2. It therefore *only depends on the quantized charges of the solution* and not on the boundary values $\phi_0 \equiv (\phi^r_0)$ of the scalar fields. This is the essence of the *attractor mechanism* [21]: The scalars along the non-flat directions of the potential V (namely which are non-trivially coupled to the black hole) flow from their values at radial infinity ϕ_0 towards fixed values at the horizon ϕ_*, solution to Eq. (4.42) and only depending on the quantized charges.

Regular extremal black holes (also called *large* extremal black holes) are then completely defined by the values of the electric and magnetic charges e, m and by boundary values ϕ_0 of the scalar fields. We shall denote such solution by $U(\tau, \phi_0), \phi^r(\tau, \phi_0)$.

---

[7]The proper distance from the horizon is measured by the variable ω such that $e^{-2U} dr^2 = d\omega^2$. Near the horizon $\omega = \log(r) = \log(-1/\tau)$ and requiring $\lim_{\omega \to -\infty} \phi^r(\omega) = \phi^r_* < \infty$ we have $\lim_{\omega \to -\infty} \frac{d^k}{d\omega^k} \phi^r = 0, k = 1, 2, \ldots,$ from which (4.41) follow.

Equations 4.32 and 4.33 can be derived from the effective action:

$$S_{eff} = \int \mathcal{L}_{eff}\, d\tau = \int \left(\dot{U}^2 + \frac{1}{2}\, G_{rs}(\phi)\, \dot{\phi}^r\, \dot{\phi}^s + e^{2U}\, V(\phi; \Gamma) \right) d\tau \,, \quad (4.44)$$

together with the Hamiltonian constraint:

$$\mathcal{H} = \dot{U}^2 + \frac{1}{2}\, G_{rs}(\phi)\, \dot{\phi}^r\, \dot{\phi}^s - e^{2U}\, V(\phi; \Gamma) = c^2 \,. \quad (4.45)$$

We see that the radial evolution of the scalar fields $\phi^r(\tau)$ and of the warp function $U(\tau)$ is described by an autonomous Hamiltonian system in which τ plays the role of time and c^2 of the conserved energy. We are interested in those solutions for which $c^2 \geq 0$. Note that in the analogy with ordinary Hamiltonian mechanics, the system is characterized by a negative definite potential $-e^{2U}\, V(\phi; \Gamma) < 0$, much like a Kepler system with conserved "energy" c^2.[8]

4.3.1 Radial Evolution from an Autonomous Hamiltonian System

Just as for a mechanical system, let us treat the warp function and the scalar fields as generalized coordinates $q^i(\tau) = (U(\tau),\, \phi^r(\tau))$, $i = 1, \ldots, n = n_s + 1$, and rewrite the effective lagrangian in the following form

$$\mathcal{L}_{eff} = \frac{1}{2}\, G_{ij}(q)\, \dot{q}^i\, \dot{q}^j + \mathcal{V}(q) \,, \quad (4.46)$$

where we have defined the metric in the "kinetic energy" term as follows

$$G_{ij}(q) = \begin{pmatrix} 2 & 0 \\ 0 & G_{rs}(\phi) \end{pmatrix} \,, \quad (4.47)$$

and $\mathcal{V}(q) \equiv e^{2U}\, V(\phi)$. Next we move to the Hamiltonian formalism by introducing conjugate momenta p_i

$$p_i = \frac{\delta \mathcal{L}_{eff}}{\delta \dot{q}^i} = G_{ij}\, \dot{q}^j \,. \quad (4.48)$$

---

[8]That with the Kepler problem is an instructive, though only qualitative, analogy. The non-extremal solutions $c^2 > 0$ correspond to elliptical (positive energy) trajectories, while the extremal ones $c^2 = 0$ to parabolic trajectories.

In terms of the phase-space variables q^i and p_i the Hamiltonian $\mathcal{H}(p, q)$ then reads:

$$\mathcal{H}(p, q) = \frac{1}{2} p_i \, G^{ij} \, p_j - \mathcal{V}(q) = \frac{1}{2} \dot{q}^i \, G_{ij}(q) \, \dot{q}^j - \mathcal{V}(q) \,. \qquad (4.49)$$

This is consistent with the constraint (4.34) that acquires the meaning of "energy conservation":

$$\mathcal{H}(p, q) = c^2 \quad \Leftrightarrow \quad \frac{1}{2} \dot{q}^i \, G_{ij}(q) \, \dot{q}^j - \mathcal{V}(q) = c^2 \,. \qquad (4.50)$$

Let us recall how the solutions to the equations of motion can be obtained by applying the machinery of the Hamilton–Jacobi theory [6]. We consider the principal Hamiltonian function $S(q, P, \tau)$ depending on q^i and on new constant momenta P_i. It is defined by the set of first order equations:

$$\frac{\partial S}{\partial q^i} = p_i \ , \quad \frac{\partial S}{\partial P_i} = Q^i \ , \quad \frac{\partial S}{\partial \tau} = -\mathcal{H} \,, \qquad (4.51)$$

where P_i, Q^i are new constant canonical variables which can be expressed in terms of the initial values of q^i and p_i. From the general theory of canonical transformations, see for instance [7], it is known that the above transformation generated by S always exists locally in the p, q space, in a neighborhood of any point which is not critical, namely in which $(\frac{\partial \mathcal{H}}{\partial q}, \frac{\partial \mathcal{H}}{\partial p}) \neq (\mathbf{0, 0})$.

We shall leave the dependence of S on the constant P_i implicit and focus on its dependence on the q^i.

From the last and the first equations of (4.51) and from the Hamiltonian constraint (4.50) we have:

$$S(q, \tau) = W(q) - c^2 \tau \ ; \quad p_i = \frac{\partial W}{\partial q^i} \,, \qquad (4.52)$$

$$\mathcal{H}(q, \partial_q W) \equiv \frac{1}{2} \partial_i W \, G^{ij} \, \partial_j W - \mathcal{V}(q) = c^2 \,, \qquad (4.53)$$

where (4.53) defines the Hamilton–Jacobi equation for the function W, usually called Hamilton characteristic function.

From Eqs. (4.52) and (4.48) we finally get

$$\dot{q}^i = G^{ij} \, p_j = G^{ij} \, \frac{\partial W}{\partial q^j} \,. \qquad (4.54)$$

Whenever $W = W(q, P; \Gamma)$ exists, the radial evolution of the scalar fields and the warp function is governed by a first order "gradient-flow" system of equations. The existence of the general solution to the Hamilton-Jacobi problem posed above is strictly related to the Liouville-integrability of the model, since the quantities P_i

provide n prime integrals in involution. In this respect a special role is played by those supergravity models which have a homogeneous symmetric scalar manifold (symmetric models), since these were shown in [17] to be related to Liouville-integrable systems. We would like to stress at this point that we are *not* interested in the most general solution W to the Hamilton-Jacobi equations. This would just mean to describe the totality of the solutions to the second order equations. We shall be interested only in the W-function associated with specific classes of solutions: the regular black holes. Under very general requirements, as we shall show below for the extremal solutions, such function *exists* and can be given in an integral form [6].

Locally, in the neighborhood of a point $\tilde{q} \equiv (\tilde{q}^i)$ we can trade the integration constants P_i with $\tilde{q}^i$ and write the general solution W to the Hamilton-Jacobi equation in the form (see for instance [7])

$$W(q_0, \tilde{q}) = W(\tilde{q}) + \int_{\tilde{\tau}, \tilde{q}}^{\tau_0, q_0} \left[c^2 + \mathcal{L}(q, \dot{q}) \right] d\tau = W(\tilde{q}) + 2 \int_{\tilde{\tau}, \tilde{q}}^{\tau_0, q_0} \left[c^2 + \mathcal{V}(q) \right] d\tau,$$
(4.55)

where we have used the Hamiltonian constraint and the integral is performed along the characteristic trajectory $\gamma = (q^i(\tau))$, i.e. the solution of Hamilton's equations, such that:

$$q^i(\tilde{\tau}) = \tilde{q}^i \ , \quad q^i(\tau_0) = q_0^i.$$
(4.56)

Example 1. Let us review the construction of W for the Reissner-Nordström black hole [32]. The q^i variables now consist of the function U alone. This is for instance a solution to $N = 2$ pure supergravity. With respect to the only vector field of the theory (the graviphoton) the solution can have in general an electric and a magnetic charge e, m. The geodesic potential reads:

$$\mathcal{V}(U, e, m) = e^{2U} Q^2 \ , \quad Q^2 \equiv \frac{1}{2}(e^2 + m^2).$$
(4.57)

The Hamiltonian constraint and the Hamilton–Jacobi equation read:

$$\dot{U}^2 = (\partial_U W)^2 = c^2 + e^{2U} Q^2.$$
(4.58)

We can then readily apply Eq. (4.55) to find, upon changing variables from τ to U:

$$
\begin{aligned}
W(U) &= W_0 + 2 \int_{U_0, \tau_0}^{U, \tau} \left[c^2 + e^{2U} Q^2 \right] d\tau = W_0 + 2 \int_{U_0}^{U} \left[c^2 + e^{2U} Q^2 \right] \frac{dU}{\dot{U}} \\
&= W_0 + 2 \int_{U_0}^{U} \sqrt{c^2 + e^{2U} Q^2} \, dU \\
&= W_0 + 2 \left[\sqrt{c^2 + e^{2U} Q^2} - \frac{c}{2} \log \left(\frac{\sqrt{c^2 + e^{2U} Q^2} + c}{\sqrt{c^2 + e^{2U} Q^2} - c} \right) \right].
\end{aligned}
$$
(4.59)

4.3.2   Critical Points as Attractors

The first order system (4.54) may have *equilibrium points* in the space of the $q^i s$, namely points (q^i_*) in which its right hand side vanishes: $\partial_i W_{|q*} = 0$. From the Hamilton–Jacobi equation it is clear that such points may exist only in the extremal case, since the right hand side of the equation

$$\partial_i W G^{ij} \partial_j W = 2 (c^2 + e^{2U} V(\phi)), \tag{4.60}$$

being $V \geq 0$, may vanish only if $c = 0$. In the extremal case we can make the ansatz $W(U, \phi^r) = 2 e^U W(\phi^r)$ and the Hamilton–Jacobi equation translates in the following equation for $W(\phi^r)$:

$$W^2 + 2 G^{rs} \frac{\partial W}{\partial \phi^r} \frac{\partial W}{\partial \phi^s} = V(\phi^r). \tag{4.61}$$

The function $W(\phi^r; \Gamma)$ is known in the literature as the *fake-superpotential* [15]. In terms of it the first order flow-equations read:

$$\dot{U} = e^U W, \qquad \dot{\phi}^r = 2 e^U G^{rs} \partial_s W, \tag{4.62}$$

Notice that the horizon points ϕ_* defined by Eq. (4.42) and reached by extremal solutions in the limit $\tau \to -\infty$, are equilibrium points of both systems (4.54) and (4.62), by virtue of Eqs. (4.40) and (4.41). In particular

$$\left. \frac{\partial W}{\partial \phi^r} \right|_{\phi_*} = 0. \tag{4.63}$$

As we shall prove below, the W-function for extremal solutions have the same symmetry properties of V and thus the same *flat-directions*. The value of W at its extremum ϕ_* can be read from (4.61) and coincides with $V_{ex}(e, m)$, namely with the horizon area:

$$W(\phi_*; e, m) = V_{ex}(e, m) = \frac{A_H(e, m)}{4\pi}. \tag{4.64}$$

From the first of Eq. (4.62) we find that W computed on the solution at radial infinity coincides with the ADM mass:

$$W(\phi_0; \Gamma) = M_{ADM}(\phi_0; \Gamma), \tag{4.65}$$

where ϕ^r_0 are the values of the scalar fields at radial infinity.

Consider now regular extremal solutions $\phi^r(\tau, \phi_0, U_0)$, $U(\tau, \phi_0, U_0)$, defined by a given vector of electric and magnetic charges Γ and by the boundary values of the fields at a generic $\tau = \tau_0 \leq 0$ (not necessarily corresponding to the radial infinity). The scalar fields $\phi^r(\tau, \phi_0, U_0)$ and the warp function $U(\tau, \phi_0, U_0)$ of the solution flow from the values ϕ_0^r, U_0 at τ_0 to ϕ_*^r at $\tau \to \tau_* = -\infty$, where ϕ_*^r satisfy both Eqs. (4.42) and (4.63) for the given electric-magnetic charges and $U_* = \lim_{\tau \to -\infty} U = -\infty$. As previously emphasized, the point ϕ_* is not uniquely expressed in terms of Γ, since both V and W in general admit a locus of flat directions (ϕ^a), $a = 1, \dots, n_f$, depending on Γ. It is however uniquely defined through the flow $\phi^r(\tau, \phi_0)$ by the boundary values ϕ_0^r of the scalar fields. We can therefore write in general $\phi_*^r = \phi_*^r(\phi_0; \Gamma)$, where only the values ϕ_*^a along the flat directions of the potential V may depend on ϕ_0.

Let us associate with such class of solutions a W-function by using the general Eq. (4.55) in which we take $\tilde{\tau} = \tau_* = -\infty$ and $\tilde{q} = \{U_*, \phi_*^r\}$. The characteristic curves are just the extremal solutions $\phi^r(\tau, \phi_0, U_0)$, $U(\tau, \phi_0, U_0)$. Since $\phi_*^r = \phi_*^r(\phi_0; \Gamma)$, we shall use the short-hand notation $W(\phi_0; \Gamma) = W(\phi_0, \phi_*(\phi_0; \Gamma); \Gamma)$ and write [6]:

$$e^{U_0} W(\phi_0; \Gamma) = e^{U_*} W(\phi_*; \Gamma) + \int_{\tau_*}^{\tau_0} e^{2U(\tau; U_0, \phi_0)} V(\phi(\tau; U_0, \phi_0); \Gamma) \, d\tau , \quad (4.66)$$

Note that the general properties of our class of solutions imply the following boundary condition for W: $e^{U_*} W(\phi_*; \Gamma) = e^{U_*} V_{ex}(e, m) = 0$. We can directly check that Eq. (4.66) yields Eq. (4.62). For the sake of simplicity we shall omit the dependence on τ and on the boundary values of the fields in the integrand. Since the integral is computed along solutions, we can use the Hamiltonian constraint (4.45) to rewrite W as follows:

$$e^{U_0} W(\phi_0; \Gamma) = e^{U_*} W(\phi_*; \Gamma) + \frac{1}{2} \int_{\tau_*}^{\tau_0} \left[e^{2U} V(\phi; \Gamma) + \dot{U}^2 + \frac{1}{2} G_{rs} \dot{\phi}^r \dot{\phi}^s \right] d\tau$$

$$= e^{U_*} W(\phi_*; \Gamma) + \frac{1}{2} \int_{\tau_*}^{\tau_0} \mathcal{L}_{eff}(U, \phi, \dot{U}, \dot{\phi}) \, d\tau . \quad (4.67)$$

Let us perform an infinitesimal variation of the boundary conditions: $U_0 \to U_0 + \delta U_0$ and $\phi_0 = \phi_0 + \delta\phi_0$. This will determine a new solution within the same class:

$$U(\tau; \phi_0 + \delta\phi_0, U_0 + \delta U_0) = U(\tau; \phi_0, U_0) + \delta U(\tau) ,$$

$$\phi(\tau; \phi_0 + \delta\phi_0, U_0 + \delta U_0) = \phi(\tau; \phi_0, U_0) + \delta\phi(\tau) . \quad (4.68)$$

Varying both sides of (4.67) under (4.68), integrating by parts and using the equations of motion we find:

$$\delta U_0 \, e^{U_0} W(\phi_0; \Gamma) + e^{U_0} \partial_r W(\phi_0; \Gamma) \delta\phi_0^r = \delta(e^{U_*} W(\phi_*; \Gamma))$$

$$+ \frac{1}{2} \int_{\tau_*}^{\tau_0} \left[\left(\frac{\partial}{\partial U} \mathcal{L}_{eff} - \frac{d}{d\tau} \frac{\partial}{\partial \dot{U}} \mathcal{L}_{eff} \right) \delta U + \left(\frac{\partial}{\partial \phi^r} \mathcal{L}_{eff} - \frac{d}{d\tau} \frac{\partial}{\partial \dot{\phi}^r} \mathcal{L}_{eff} \right) \delta\phi^r \right]$$

$$+ \left. \left(\dot{U} \delta U + \frac{1}{2} G_{rs} \dot{\phi}^s \delta\phi^r \right) \right|_{\tau_*}^{\tau_0} = \delta(e^{U_*} W(\phi_*; \Gamma)) + \left. \left(\dot{U} \delta U + \frac{1}{2} G_{rs} \dot{\phi}^s \delta\phi^r \right) \right|_{\tau_*}^{\tau_0} .$$

$$(4.69)$$

Having chosen $\tau_* = -\infty$, all terms computed at τ_* in the above equation vanish by virtue of the near horizon behavior of the regular extremal solutions. Equating the variations at τ_0 on both sides we find:

$$\dot{U}(\tau_0) = e^{U_0}\, W(\phi_0; \Gamma) \;,\quad \dot{\phi}^s(\tau_0) = 2\, e^{U_0}\, G^{rs}(\phi_0)\, \partial_r W(\phi_0; \Gamma)\,. \quad (4.70)$$

Being τ_0 generic, we find that W defines the first order Eq. (4.62) for the fields and thus it is a solution to the Hamilton-Jacobi equation.

Note however that W, as defined in (4.66), may in principle depend on the chosen value of τ_0, that is $W = W(U_0, \phi_0, \tau_0; \Gamma)$. Let us show that this is not the case, namely that $W(U_0, \phi_0, \tau_0 + \delta\tau; \Gamma) = W(U_0, \phi_0, \tau_0; \Gamma)$, for a generic $\delta\tau$. To do this we vary $\tau_0 \to \tau_0 + \delta\tau$, *keeping the boundary values of the fields fixed*. This requires to change the solution on which the integral is computed from $U(\tau), \phi(\tau)$ to $U'(\tau), \phi'(\tau)$ such that:

$$U'(\tau_0 + \delta\tau) = U(\tau_0) = U(\tau_0 + \delta\tau) - \dot{U}(\tau_0)\,\delta\tau\,,$$
$$\phi'(\tau_0 + \delta\tau) = \phi(\tau_0) = \phi(\tau_0 + \delta\tau) - \dot{\phi}(\tau_0)\,\delta\tau\,. \quad (4.71)$$

and thus amounts to performing, along the flow, the transformation $U \to U - \dot{U}\,\delta\tau$, $\phi \to \phi - \dot{\phi}\,\delta\tau$, besides changing the domain of integration, $\delta\tau$ being chosen along the flow so that $\delta\tau_* = 0$. After some straightforward calculations we find:

$$e^{U_0}\,(W(U_0, \phi_0, \tau_0 + \delta\tau; \Gamma) - W(U_0, \phi_0, \tau_0; \Gamma)) = H|_{\tau_0}\,\delta\tau = 0\,, \quad (4.72)$$

in virtue of the Hamiltonian constraint. Since the function W of the moduli space, as defined by (4.66), does not depend on the choice of τ_0, we can choose $\tau_0 = 0$, where $U_0 = 0$ and then find:

$$W(\phi_0; \Gamma) = \int_{-\infty}^{0} e^{2\,U(\tau;\phi_0)}\, V(\phi(\tau;\phi_0); \Gamma)\,d\tau\,. \quad (4.73)$$

Note that, being V positive definite, W is positive definite as well. Moreover if we compute W on a solution and define $W(\tau) \equiv W(\phi(\tau))$, its τ-derivative is positive:

$$\frac{d}{d\tau} W(\tau) = \dot{\phi}^r\,\partial_r W = 2G^{rs}\partial_r W\,\partial_s W > 0 \qquad \tau > -\infty\,, \quad (4.74)$$

Therefore $W(\tau)$ monotonically increases from its value $A_H(e, m)/(4\pi)$ at the horizon to $M_{ADM}(\phi_0; e, m)$ at radial infinity. The superpotential W has the properties of a Liapunov function [6][9] which implies that ϕ_* is an *asymptotically stable equilibrium* point (or *stable attractor point*) for the scalar dynamical system

---

[9]See for instance [29] for a definition of stability according to Liapunov and of a Liapunov's function.

in (4.62). In general the very existence of a solution W to the Hamilton-Jacobi equation, even just in a neighborhood of its critical point ϕ_* guarantees (by Liapunov's theorem [29]) that the point is a stable attractor for the scalar fields of the corresponding solution.[10]

The correspondence illustrated above, between the class of regular extremal solutions associated with a given vector of electric-magnetic charges and a W-function, is general, namely it *holds for any supergravity model*, including those with non-homogeneous scalar manifolds, which are in general not integrable. We stress here that the W function constructed above is far from being the general solution to the Hamilton-Jacobi equation, which itself may not exist.

4.4 Duality

Let us now consider the effect of the global symmetries discussed in Sect. 4.2, on black hole solutions. A duality transformation $g \in G$ will map a black hole solution $U(\tau), \phi^r(\tau)$ with charges Γ into a new solution $U'(\tau) = U(\tau)$, $\phi'^r(\tau) = g \star \phi^r(\tau)$ with charges $\Gamma' = D(g)\Gamma$. More specifically, if $U(\tau), \phi^r(\tau)$ is defined by the boundary condition ϕ_0 for the scalar fields, $U'(\tau) = U(\tau)$, $\phi'^r(\tau)$ is the *unique solution*, within our class, with charges Γ' defined by the boundary condition $\phi_0' = g \star \phi_0$

$$g \in G : \begin{cases} U(\tau; \phi_0) \\ \phi(\tau; \phi_0) \\ \Gamma \end{cases} \longrightarrow \begin{cases} U'(\tau; g \star \phi_0) = U(\tau; \phi_0) \\ \phi'(\tau; g \star \phi_0) = g \star \phi(\tau; \phi_0) \\ \Gamma' = D(g)\Gamma \end{cases} . \quad (4.75)$$

The central and matter charges defined in Eq. (4.30), on the other hand, under a transformation (4.75) only transform with the H-compensating transformation $h(\phi^r, g)$ in (4.1). To see this let us write (4.30) in matrix notation, denoting by $\mathbf{Z}$ the vector $(Z_{\hat{M}})$ (we ignore the Cayley matrix since it just amounts to a change of basis):

$$\mathbf{Z}(\phi, \Gamma) = \mathcal{D}[\mathbb{L}(\phi)]^T \mathbb{C}\Gamma . \quad (4.76)$$

Upon a duality transformation (4.75) we find:

$$\mathbf{Z}(g \star \phi, \mathcal{D}[g]\Gamma) = \mathcal{D}[\mathbb{L}(g \star \phi)]^T \mathbb{C}\mathcal{D}[g]\Gamma = \mathcal{D}[h(g, \phi)]^{-T} \mathcal{D}[\mathbb{L}(\phi)]^T \mathcal{D}[g]^T \mathbb{C}\mathcal{D}[g]\Gamma$$

$$= \mathcal{D}[h(g, \phi)]^{-T} \mathbf{Z}(\phi, \Gamma) . \quad (4.77)$$

Using (4.16) and the definition (4.35), we see that the effective potential is invariant if we act on ϕ^r and Γ by means of G simultaneously:

$$V(\phi; \Gamma) = V(g \star \phi; D(g)\Gamma) . \quad (4.78)$$

---

[10] See the second of [6] for a discussion of subtleties related to the existence of flat directions.

This implies that V, as a function of the scalar fields and quantized charges, is G-invariant. From this property of V it follows that the effective action (4.44) and the extremality constraint (4.45) are manifestly duality invariant.

All the duality-invariant functions of the scalars and the electric-magnetic charges can be built as H-invariant functions of the central and matter charges. For instance the scalar potential $V(\phi; \Gamma)$ can be written in the following form:

$$V(\phi; \Gamma) = \frac{1}{2} \bar{Z}^{AB} Z_{AB} + \bar{Z}^I Z_I = \mathbf{Z}^\dagger \mathbf{Z}. \tag{4.79}$$

Let us show now that the W function shares with V the same symmetry property (4.78), namely that it is G-invariant as well:

$$W(\phi; \Gamma) = W(g \star \phi; D(g) \Gamma). \tag{4.80}$$

This is easily shown using the general form Eqs. (4.73) and (4.75):

$$
\begin{aligned}
W(g \star \phi_0; D(g) \Gamma) &= \int_{-\infty}^{0} e^{2 U'(\tau; g \star \phi_0)} V(\phi'(\tau; g \star \phi_0); D(g) \Gamma) \, d\tau \\
&= \int_{-\infty}^{0} e^{2 U(\tau; \phi_0)} V(g \star \phi(\tau; \phi_0); D(g) \Gamma) \, d\tau \\
&= \int_{-\infty}^{0} e^{2 U(\tau; \phi_0)} V(\phi(\tau; \phi_0); \Gamma) \, d\tau = W(\phi_0; \Gamma) \,.
\end{aligned}
\tag{4.81}
$$

Being the ADM mass expressed in terms of W, see Eq. (4.65), it is a G-invariant quantity as well:

$$M_{ADM}(\phi_0; \Gamma) = M_{ADM}(g \star \phi_0; D(g) \Gamma). \tag{4.82}$$

Extremal black-holes can be grouped into orbits with respect to the duality action (4.75) of G. These orbits are characterized in terms of G-invariant functions of the scalar fields and the quantized charges, which are expressed in terms of H-invariant functions of the central and matter charges. Expressing W as an H-invariant combination of the $Z_{\hat{M}}$ is in general a rather difficult task. It was partly accomplished in [4, 16] for various duality orbits of extremal black holes. For certain classes of solutions this invariant function is not known yet (see Sect. 4.4.2).

Let us now consider the value $V_{ex}(\Gamma)$ of the scalar potential $V(\phi; \Gamma)$ in its extremum ϕ_*. As pointed out earlier, it only depends on the electric and magnetic charges. Moreover, since both V and the Eq. (4.42) are G-invariant, $V_{ex}(\Gamma)$ should be a duality invariant function of the quantized charges only:

$$V_{ex}(\mathcal{D}[g]\Gamma) = V_{ex}(\Gamma). \tag{4.83}$$

There is only one such invariant function, and it is the *quartic invariant* $I_4(\Gamma)$ of the symplectic representation of G in which Γ transforms. It turns out that the scalar potential at its extremum can be expressed as follows:

$$V_{ex}(\Gamma) = \sqrt{|I_4(\Gamma)|}. \tag{4.84}$$

This implies, using Eq. (4.43), that the horizon area $A_H(e, m)$ of regular extremal black holes can be expressed in terms of the quartic invariant of the duality group G: $A_H(e, m) = 4\pi \sqrt{|I_4(\Gamma)|}$. In fact I_4 is the only G-invariant which characterizes the near horizon geometry of extremal black holes. An other duality invariant feature of black holes is the amount of supersymmetry they preserve. In an N-extended supergravity, regular black holes can preserve at most the Nth part of the supersymmetry of the theory. These are the BPS regular black holes, which have played an important role in the microscopic description of the black hole entropy, as recalled in the introduction. They are extremal solutions and the corresponding first order dynamical system (4.62) can be derived from the Killing spinor equations, the W function being expressed in terms of the modulus of the skew-eigenvalue of the central charge matrix Z_{AB} associated with the preserved supersymmetry. For $N = 2$ theories, the central charge matrix can be written as $Z_{AB} = Z \epsilon_{AB}$ and $W(\phi; \Gamma) = |Z(\phi, \Gamma)|$ (see below). The quartic invariant for regular BPS black holes is positive.

Extremal solutions, as anticipated in the introduction, can also be non-supersymmetric. In this case I_4 need not be positive and in fact, for a certain class of non-supersymmetric extremal solutions it is negative. A thorough classification of regular extremal black holes according to their duality-invariant near-horizon geometries was performed in [10].

Electric-magnetic charge orbits for which $I_4(\Gamma) = 0$ define extremal solutions with vanishing horizon area. These are the so called *small* black holes. They are characterized by the following power-law behavior of the warp function as $\tau \to -\infty$: $e^{-2U} \sim \tau^\alpha$ with $\alpha < 2$. Such solutions can be obtained from the regular ones by setting some of the charges to zero and the duality invariant properties associated with their electric-magnetic charges were studied in [14, 20]. For small black holes the value of the potential (and thus of the W-function) vanishes along the solution as $\tau \to -\infty$ since $V_{ex}(e, m) = 0$. Since V, defined in (4.35), is positive definite, it can be zero only in the limit in which some of the scalar fields run to infinity. This implies that the point ϕ_*^r reached by the scalar fields as $\tau \to -\infty$ belongs to the boundary of the scalar manifold.

4.4.1 The Issue of Flat Directions

Let us denote by $G_0 \subset G$ the *little group* (or stabilizer) of the orbit of the quantized charges Γ under the action of G [10, 14]:

$$g_0 \in G_0: \quad D(g_0) \Gamma = \Gamma. \tag{4.85}$$

Table 4.1 Summary of regular, extremal black hole orbits in the various supergravities. The symbols I, II, III denote the $\frac{1}{N}$-BPS, the non-BPS ($I_4 > 0$) and the non-BPS ($I_4 < 0$) orbits respectively. For those solutions with non-trivial moduli spaces $\frac{G_0}{H_0}$ (i.e. G_0 non-compact)

$\mathcal{N}$	$\dfrac{G}{H}$	Orbit	$\dfrac{G_0}{H_0}$
8	$\dfrac{E_{7(7)}}{SU(8)}$	I	$\dfrac{E_{6(2)}}{SU(2)\times SU(6)}$
		III	$\dfrac{E_{6(6)}}{USp(8)}$
6	$\dfrac{SO^*(12)}{U(6)}$	I	$\dfrac{SU(4,2)}{S[U(4)\times U(2)]}$
		II	—
		III	$\dfrac{SU^*(6)}{USp(6)}$
5	$\dfrac{SU(5,1)}{U(5)}$	I	$\dfrac{SU(2,1)}{U(2)}$
4	$\dfrac{SL(2,\mathbb{R})}{SO(2)} \times \dfrac{SO(6,n)}{SO(6)\times SO(n)}$	I	$\dfrac{SO(4,n)}{SO(4)\times SO(n)}$
		II	$\dfrac{SO(6,n-2)}{SO(6)\times SO(n-2)}$
		III	$SO(1,1) \times \dfrac{SO(5,n-1)}{SO(5)\times SO(n-1)}$
3	$\dfrac{SU(3,n)}{S[U(3)\times U(n)]}$	I	$\dfrac{SU(2,n)}{S[U(2)\times U(n)]}$
		II	$\dfrac{SU(3,n-1)}{S[U(3)\times U(n-1)]}$
		III	—
	$\dfrac{SU(1,n+1)}{U(n+1)}$	I	—
		II	$\dfrac{SU(1,n)}{U(n)}$
2	$\dfrac{SL(2,\mathbb{R})}{SO(2)} \times \dfrac{SO(2,n+2)}{SO(2)\times SO(n+2)}$	I	—
		II	$\dfrac{SO(2,n)}{SO(2)\times SO(n)}$
		III	$SO(1,1) \times \dfrac{SO(1,n+1)}{SO(n+1)}$
	$\dfrac{Sp(6)}{U(3)}$	I	—
		II	$\dfrac{SU(2,1)}{U(2)}$
		III	$\dfrac{SL(3,\mathbb{R})}{SO(3)}$
	$\dfrac{SU(3,3)}{S[U(3)\times U(3)]}$	I	—
		II	$\left(\dfrac{SU(2,1)}{U(2)}\right)^2$
		III	$\dfrac{SL(3,\mathbb{C})}{SU(3)}$
	$\dfrac{SO^*(12)}{U(6)}$	I	—
		II	$\dfrac{SU(4,2)}{S[U(4)\times U(2)]}$
		III	$\dfrac{SU^*(6)}{USp(6)}$
	$\dfrac{E_{7(-25)}}{U(1)\times E_6}$	I	—
		II	$\dfrac{E_{6(-14)}}{U(1)\times SO(10)}$
		III	$\dfrac{E_{6(-26)}}{F_4}$

Of course the embedding of G_0 within G depends in general on Γ. Let us show that the scalar fields φ^a spanning the submanifold G_0/H_0, H_0 being the maximal compact subgroup of G_0, are *flat directions* of the potential and of the W-function, namely that neither V nor W, depend on φ^a. Since we are interested in the part of the little group which has a free action on the moduli, we shall define G_0 modulo compact group-factors. For instance if the little group is $SU(3)\times SU(2,1)$, we define G_0 to be $SU(2,1)$ and thus $H_0 = U(2)$. For a summary of the orbits of regular extremal black holes in the various theories and of the corresponding moduli spaces G_0/H_0 see Table 4.1.

To prove that φ^a are flat directions of both V and W, let us decompose the n scalar fields ϕ^r into the φ^a scalars parametrizing the submanifold G_0/H_0 and scalars φ^k, which can be chosen to transform linearly with respect to H_0. Let us stress at this point that the coordinates φ^a, φ^k will in general depend on the original ones ϕ^r and on the electric and magnetic charges, namely:

$$\varphi^a = \varphi^a(\phi^r, p^\Lambda, q_\Lambda) , \quad \varphi^k = \varphi^k(\phi^r, p^\Lambda, q_\Lambda) . \tag{4.86}$$

Let us choose, for convenience, a basis of coordinates in the moduli space such that the first n_f components of ϕ^r coincide with the φ^a, the others being φ^k, that is $\phi^a = \varphi^a$, $\phi^k = \varphi^k$. We can move along the ϕ^a direction through the action of isometries in G_0. Let us consider infinitesimal isometries in G_0 whose effect is to shift the a-scalars only:

$$g_0 \in G_0 : \quad \phi^r \rightarrow (g_0 \star \phi)^r = \phi^r + \delta^r_a \, \delta\phi^a , \quad \Gamma \rightarrow \Gamma' = \Gamma + \delta\Gamma = \Gamma ,$$

where we have used the definition of G_0, (4.85). Let us now use Eqs. (4.78) and (4.80) to evaluate the corresponding infinitesimal variations of V and W:

$$V(\phi^r; \Gamma) = V(\phi^r + \delta\phi^r; \Gamma + \delta\Gamma) = V(\phi^k, \phi^a + \delta\phi^a; \Gamma) .$$
$$W(\phi^r; \Gamma) = W(\phi^r + \delta\phi^r; \Gamma + \delta\Gamma) = W(\phi^k, \phi^a + \delta\phi^a; \Gamma) . \tag{4.87}$$

We conclude that $\frac{\partial V}{\partial \phi^a} = \frac{\partial W}{\partial \phi^a} = 0$, namely that ϕ^a are flat direction of both functions. Using Eq. (4.65) we see that the same property holds for the ADM mass: $\frac{\partial}{\partial \phi^a} M_{ADM} = 0$.

Let us now give a general characterization of the W-function in terms of the central and matter charges. We can write the coset representative $\mathbb{L}(\phi^r)$ of $\mathcal{M}_{scal}$ as the product of the G_0/H_0 coset representative $\mathbb{L}_0(\phi^a)$ times a matrix $\mathbb{L}_1(\phi^k)$ depending on the remaining scalars:

$$\mathbb{L}(\phi^r) = \mathbb{L}(\phi^a, \phi^k) = \mathbb{L}_0(\phi^a) \, \mathbb{L}_1(\phi^k) . \tag{4.88}$$

Correspondingly the vector $\mathbf{Z}(\phi, \Gamma)$ of central and matter charges, as defined in (4.30) can be written as follows (as usual we omit the Cayley transformation):

$$\mathbf{Z}(\phi^r, \Gamma) = \mathcal{D}[\mathbb{L}_1(\phi^k)]^T \, \mathcal{D}[\mathbb{L}_0(\phi^a)]^T \, \mathbb{C}\,\Gamma , \tag{4.89}$$

Now we can use the property that $\mathbb{L}_0(\phi^a)$ belongs to G_0 in the symplectic representation, so that $\mathcal{D}[\mathbb{L}_0]^T \, \mathbb{C}\,\Gamma = \mathbb{C}\,\mathcal{D}[\mathbb{L}_0]^{-1}\,\Gamma = \mathbb{C}\,\Gamma$ and write:

$$\mathbf{Z}(\phi^a, \phi^k, \Gamma) = \mathcal{D}[\mathbb{L}_1(\phi^k)]^T \, \mathbb{C}\,\Gamma = \mathbf{Z}(0, \phi^k, \Gamma) , \tag{4.90}$$

that is the central and matter charges do not depend on ϕ^a at all:

$$\frac{\partial}{\partial \phi^a} Z_{AB} = \frac{\partial}{\partial \phi^a} Z_I = 0 . \tag{4.91}$$

Let us now describe the effect of a generic transformation g_0 in G_0 on the central charges. From the general properties of coset representatives we know that $g_0 \mathbb{L}_0(\phi^a) = \mathbb{L}_0(g_0 \star \phi^a) h_0$, h_0 being a compensator in H_0 depending on g_0 and ϕ^a. Now, using the property that ϕ^k transform in a linear representation of H_0, we can describe the action of g_0 on a generic point ϕ as follows:

$$g_0 \mathbb{L}(\phi^r) = g_0 \mathbb{L}_0(\phi^a) \mathbb{L}_1(\phi^k) = \mathbb{L}_0(g_0 \star \phi^a) h_0 \mathbb{L}_1(\phi^k) h_0^{-1} h_0$$
$$= \mathbb{L}_0(g_0 \star \phi^a) \mathbb{L}_1(\phi'^k) h_0 = \mathbb{L}(g_0 \star \phi^r) h_0 , \tag{4.92}$$

where ϕ'^k is the transformed of ϕ^k by h_0, and $(g_0 \star \phi^a, \phi'^k)$ define the transformed $g_0 \star \phi^r$ of ϕ^r by g_0. From (4.92) and the definition (4.30) we derive the following property:

$$\forall g_0 \in G_0 : \mathbf{Z}(g_0 \star \phi^r, \Gamma) = \mathcal{D}[h_0]^{-T} \mathbf{Z}(\phi^r, \Gamma) . \tag{4.93}$$

Now consider the W function as a function of ϕ^r and Γ through the central and matter charges $Z_{\hat{M}}$:

$$W(\phi^r; \Gamma) = \widehat{W}[\mathbf{Z}(\phi^r, \Gamma)] . \tag{4.94}$$

From the duality-invariance of W it follows that, for any $g_0 \in G_0$ we have

$$W(\phi^r; \Gamma) = W(g_0 \star \phi^r; \mathcal{D}[g_0] \Gamma) = W(g_0 \star \phi^r; \Gamma) . \tag{4.95}$$

Furthermore, using Eqs. (4.93) and (4.94) we find:

$$\widehat{W}[\mathbf{Z}(\phi^r, \Gamma)] = W(\phi^r; \Gamma) = W(g_0 \star \phi^r; \Gamma) = \widehat{W}[\mathbf{Z}(g_0 \star \phi^r, \Gamma)]$$
$$= \widehat{W}[\mathcal{D}[h_0]^{-T} \mathbf{Z}(\phi^r, \Gamma)] . \tag{4.96}$$

The above equality holds for any $g_0 \in G_0$ and thus for any $h_0 \in H_0$. We conclude from this that W *can be characterized, for a given orbit of solutions, as an H_0-invariant function of the central and matter charges.* Let us stress once more that we have started from a generic charge vector Γ, so that the definition of G_0, and thus of H_0, is charge dependent. We could have started from a given G_0 inside G and worked out the representative Γ_0 of the G-orbit having G_0 as manifest little group. In this case, by construction, the (ϕ^a, ϕ^k) parametrization is charge-independent.

4.4.2   The W-Function for the $I_4 < 0$ Orbit

As mentioned earlier, in the case of symmetric models, the W superpotential has been found for almost all duality orbits (of extremal solutions), with the important exception of the non-supersymmetric solutions characterized by $I_4 < 0$ (solutions III of Table 4.1). The generic representative of this class is in principle known, since the corresponding *generating solution*, that is the simplest solution[11] capturing all the duality invariant features of the most general one, has been constructed in [27]. Knowing the generic representative of this duality orbit $\phi^r(\tau, \phi_0)$, $U(\tau, \phi_0)$ for a given charge vector Γ, the general Eq. (4.73) provides the corresponding (duality invariant) W-superpotential as a free function of the point ϕ_0 on the scalar manifold and of the charges.[12] Such expression is however not handy since it would be given in terms of an integral in τ from $-\infty$ to 0 of a function of τ, ϕ_0 and Γ. In [16], the explicit form of W as a free function of the scalar fields and Γ, was given in a specific $N = 2$ model, the t^3-model, characterized by a single complex scalar t spanning the manifold $SL(2, \mathbb{R})/SO(2)$. Its expression in terms of duality-invariant quantities was also given. In a generic $N = 2$ supergravity however, such solutions are characterized by $n_f = n_V - 2$ flat directions and the explicit form of W is still missing.

A partial answer to the problem was given in [16] where a certain (not duality-invariant) subset of these black holes was described in terms of a solution W_{α^a} to the Hamilton-Jacobi equation, where the parameters (α^a), $a = 1, \ldots, n_f$, define such subset. Let us comment on relation between this function and W in the light of our discussion in Sect. 4.3.2 on the solutions to the Hamilton-Jacobi equations. In a generic $N = 2$ symmetric model, given a vector Γ of electric and magnetic charges in the orbit $I_4(\Gamma) < 0$, the scalar potential $V(\phi; \Gamma)$ admits a submanifold $\mathcal{M}_c \sim G_0/H_0$ of critical points of dimension $n_f = n_V - 2$, see Table 4.1. The corresponding regular black hole solution $\phi^r(\tau, \phi_0)$ interpolates between a point $\phi_0 = (\phi_0^r)$ at radial infinity and a point ϕ_* in $\mathcal{M}_c$ which is uniquely determined by the electric-magnetic charges and by ϕ_0 (the dependence on ϕ_0 is due to the presence of flat directions): $\phi_*^r = \phi_*^r(\phi_0; \Gamma)$. The corresponding superpotential $W(\phi; \Gamma)$ describing such set of solutions is computed by evaluating the integral on the right hand side of Eq. (4.73) on such solutions $\phi^r(\tau, \phi_0)$, $U(\tau, \phi_0)$.

We could fix a point $\phi_*^r(\alpha^a)$ on $\mathcal{M}_c$ and consider the set of all solutions $\phi^{\prime r}(\tau, \phi_0)$, $U'(\tau, \phi_0)$ to the *second order field equations* for the same electric and magnetic charges interpolating between a generic point ϕ_0 at radial infinity and the chosen point $\phi_*^r(\alpha^a)$. Clearly only a subset of such solutions will describe regular black holes, namely those originating at spatial infinity from the submanifold $\mathcal{M}_{\alpha^a}$ of points ϕ_0 satisfying the condition:

$$\phi_*^r(\phi_0; \Gamma) = \phi_*^r(\alpha^a). \tag{4.97}$$

---

[11]By simplest here we mean depending on the least number of independent parameters.

[12]Notice that the electric and magnetic charges also enter the functional form of the solution $\phi^r(\tau, \phi_0)$, $U(\tau, \phi_0)$ through harmonic functions.

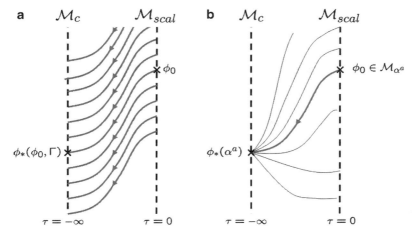

Fig. 4.1 Figurative description of the difference between W and W_{α^a}: the former (**a**) is evaluated as the integral (4.73) over the full set of regular *black holes* pertaining to the chosen set of electric and magnetic charges (*thick, red lines* in **a**); the latter as the same integral over the set of solutions (**b**) interpolating a generic point ϕ_0 with a fixed $\phi_*(\alpha^a)$ in $\mathcal{M}_c$, of which only a subset describes regular *black holes* (*thick, red line*)

The W_{α^a} function is obtained by evaluating the integral in Eq. (4.55) over this class of solutions, which means choosing $\tau_0 = 0$, $U_0 = 0$, $\tilde{\tau} = \tau_* = -\infty$, $\tilde{U} = U_* = -\infty$, but now $\tilde{\phi}^r = \phi_*^r(\alpha^a)$. In formulas we can write:

$$W(\phi_0; \Gamma) = W(\phi_0, \phi_*(\phi_0; \Gamma); \Gamma) = \int_{-\infty, \phi_*(\phi_0; \Gamma)}^{0, \phi_0} e^{2U(\tau;\phi_0)} V(\phi(\tau;\phi_0); \Gamma) d\tau,$$

$$W_{\alpha^a}(\phi_0; \Gamma) \equiv W(\phi_0, \phi_*(\alpha^a); \Gamma) = \int_{-\infty, \phi_*(\alpha^a)}^{0, \phi_0} e^{2U'(\tau,\phi_0)} V(\phi'(\tau, \phi_0); \Gamma) d\tau.$$

$$(4.98)$$

The difference between the two solutions W and W_{α^a} to the Hamilton-Jacobi equation is figuratively described in Fig. 4.1.

The class of solutions $\phi'^r(\tau, \phi_0)$, $U'(\tau, \phi_0)$ interpolating between a generic point ϕ_0 on the scalar manifold and $\phi_*^r(\alpha^a)$ is clearly not invariant under duality transformations since $\phi_*^r(\alpha^a)$ is not. In fact, by a derivation analogous to the one used in Sect. 4.4 to prove the duality invariance of W, one can easily show that, for any $g \in G$:

$$W(\phi_0, \phi_*(\alpha^a); \Gamma) = W(g \star \phi_0, g \star \phi_*(\alpha^a); \mathcal{D}[g]\,\Gamma).\qquad(4.99)$$

In [16] it was suggested that the W-function associated with the regular black holes of the $I_4 < 0$ orbit, can be obtained by extremising W_{α^a} with respect to the parameters α^a, seen as *auxiliary fields*

$$W(\phi; \Gamma) = W_{\tilde{\alpha}^a}(\phi; \Gamma) \; ; \quad \left. \frac{\partial W_{\alpha^a}}{\partial \alpha^b} \right|_{\tilde{\alpha}^a} = 0 \,. \tag{4.100}$$

The search for the W-function associated with the $I_4 < 0$ orbit, as an analytic function of the scalar fields and the charges, is still an open problem. In [12] W^2 was characterized as a zero of a degree-6 polynomial, suggesting that an algebraic expression for W may not even exist.

4.5 Time-Reductions to $D = 3$

It is known, since the pioneering paper by Breitenlohner et al. [13], that stationary solutions in $D = 4$ supergravity can be described within an Euclidean $D = 3$ supergravity which is *formally* obtained by dimensional reduction along the time-direction of the four-dimensional one (and further dualization of all the vector fields to scalars). The advantage of this approach is that the global symmetry group G_3 of the resulting lower-dimensional model is sensibly greater than the one in four-dimensions. In fact G_3 contains the product of G and of the Ehlers group $\mathrm{SL}(2, \mathbb{R})_E$ which acts transitively on the moduli arising from the four-dimensional metric. Static (shperically symmetric) solutions are described by *geodesics* on the scalar manifold $\mathcal{M}^{(3)}_{scal}$ of the three-dimensional theory.

Let us briefly review this approach and the definition of the Euclidean three-dimensional theory. To implement the reduction along the time direction, let us start from the following general ansatz for a stationary metric [13][13]:

$$ds_4^2 = -e^{2U} (dt + A_i^0 \, dx^i)^2 + e^{-2U} g_{ij}^{(3)} \, dx^i \, dx^j \,, \tag{4.101}$$

where $x^i, i = 1, 2, 3$, are the coordinates of the final Euclidean space and $g_{ij}^{(3)}$ is the corresponding positive defined metric. Denoting the time-components of the vector fields A_μ^Λ by ζ^Λ, we can define the dimensionally reduced vectors $\tilde{A}^\Lambda = \tilde{A}_i^\Lambda \, dx^i$ as follows:

$$A^\Lambda = A_\mu^\Lambda \, dx^\mu = \tilde{A}^\Lambda + \zeta^\Lambda (dt + A_i^0 \, dx^i) \,. \tag{4.102}$$

The vector field-strengths decompose correspondingly:

$$F^\Lambda = \tilde{F}^\Lambda + d\zeta^\Lambda \wedge (dt + A_i^0 \, dx^i) \,, \tag{4.103}$$

where

$$\tilde{F}_{ij}^\Lambda \equiv \partial_i \tilde{A}_j^\Lambda - \partial_j \tilde{A}_i^\Lambda + \zeta^\Lambda F_{ij}^0 \; ; \quad F_{ij}^0 \equiv \partial_i A_j^0 - \partial_j A_i^0 \,. \tag{4.104}$$

---

[13]This metric describes stationary solutions, in the static case $A_i^0 = 0$.

Using the property $\sqrt{|g|} = e^{-2U} \sqrt{|g^{(3)}|}$ and replacing the above ansätze for the fields into the four dimensional Lagrangian we find

$$\sqrt{|g|}\, \mathcal{L}_4 = \sqrt{|g^{(3)}|}\, \mathcal{L}_3' = \sqrt{|g^{(3)}|}\, \left[\frac{1}{2}\, R[g^{(3)}] - \partial_i U \partial^i U - \frac{1}{2}\, G_{rs}\, \partial_i \phi^r \partial^i \phi^s \right.$$

$$+ \frac{e^{4U}}{8}\, F_{ij}^0 F^{0\,ij} + \frac{e^{2U}}{4}\, \tilde{F}_{ij}^\Lambda I_{\Lambda\Sigma}\, F^{\Sigma\,ij} + \frac{1}{2\sqrt{|g^{(3)}|}}\, \tilde{F}_{ij}^\Lambda I_{\Lambda\Sigma}\partial_k \zeta^\Sigma$$

$$\left. - \frac{e^{-2U}}{2}\, \partial_i \zeta^\Lambda I_{\Lambda\Sigma}\, \partial^i \zeta^\Sigma \right] . \tag{4.105}$$

Next we dualize the three-dimensional vector fields $\tilde{A}_i^\Lambda$, A_i^0 into $n_V + 1$ scalar fields $\tilde{\zeta}_\Lambda$, $\tilde{a}$. This is effected by adding to the Lagrangian $\mathcal{L}_3'$ the following Chern-Simons terms[14]:

$$\sqrt{|g^{(3)}|}\, \mathcal{L}_3 = \sqrt{|g^{(3)}|}\, \mathcal{L}_3' + \epsilon^{ijk} \left[-\partial_i \tilde{A}_j^\Lambda \partial_k \tilde{\zeta}_\Lambda + \frac{1}{2}\, \partial_i A_j^0\, \partial_k \tilde{a} \right] . \tag{4.106}$$

We now extremize the above Lagrangian with respect to the vector field-strengths and find:

$$\frac{\partial \mathcal{L}_3}{\partial \tilde{F}_{ij}^\Lambda} = 0 \;\Rightarrow\; \tilde{F}^{\Lambda\,ij} = \frac{e^{-2U}}{\sqrt{|g^{(3)}|}}\, \epsilon^{ijk}\, I^{-1\,\Lambda\Sigma} \left(\partial_k \tilde{\zeta}_\Sigma - R_{\Sigma\Gamma}\partial_k \zeta^\Gamma \right) ,$$

$$\frac{\partial \mathcal{L}_3}{\partial F_{ij}^0} = 0 \;\Rightarrow\; F^{0\,ij} = -\frac{e^{-4U}}{\sqrt{|g^{(3)}|}}\, \epsilon^{ijk}\, \omega_k , \tag{4.107}$$

where

$$\omega_k \equiv \partial_k \tilde{a} + 2\, \zeta^\Lambda \partial_k \tilde{\zeta}_\Lambda = \partial_k a + \zeta^\Lambda \partial_k \tilde{\zeta}_\Lambda - \tilde{\zeta}_\Lambda \partial_k \zeta^\Lambda = \partial_k a + \mathcal{Z}^T \mathbb{C} \partial_k \mathcal{Z} , \tag{4.108}$$

and we have defined the scalar a and the symplectic vector of electric and magnetic potentials $\mathcal{Z}$ as follows:

$$\mathcal{Z} = (\mathcal{Z}^M) \equiv \begin{pmatrix} \zeta^\Lambda \\ \tilde{\zeta}_\Lambda \end{pmatrix} , \quad \tilde{a} \equiv a - \zeta^\Lambda \tilde{\zeta}_\Lambda . \tag{4.109}$$

Replacing (4.107) into $\mathcal{L}_3$ we end up with a $D = 3$ action of the form

$$S_3 = \int d^3 x \sqrt{|g^{(3)}|} \left(\frac{1}{2}\, R_3 - \frac{1}{2}\, G_{IJ}(\Phi)\, \partial_i \Phi^I\, \partial^i \Phi^J \right) ,$$

---

[14]In our conventions $\epsilon_{123} = \epsilon^{123} = +1$.

which describes a sigma-model coupled to gravity. The $n_s + 2n_V + 2$ scalar fields of the model are $\Phi^I \equiv (U, \phi^r, a, Z^M)$ and span a scalar manifold $\mathcal{M}_3$ on which the following metric is defined:

$$G_{IJ}(\Phi)\, d\Phi^I\, d\Phi^J = 2dU^2 + G_{rs}\, d\phi^r\, d\phi^s + \frac{e^{-4U}}{2}\, \omega^2 + e^{-2U}\, dZ^T\, \mathcal{M}\, dZ,$$

where $\mathcal{M} = (\mathcal{M}_{MN})$ is the symmetric, symplectic matrix defined in (4.13). Being $\mathcal{M}(\phi^r)$ negative definite, we see that the metric on $\mathcal{M}_{scal}^{(3)}$ has indefinite signature, with $2n_V$ negative-signature directions, namely $\mathcal{M}_{scal}^{(3)}$ is pseudo-Riemannian. The manifold $\mathcal{M}_{scal}^{(3)}$ contains the scalar manifold in $\mathcal{M}_{scal}$ in the $D = 4$ model as a submanifold and is homogeneous (symmetric) if and only if its four dimensional counterpart is. In particular if $\mathcal{M}_{scal}$ is homogeneous symmetric of the form G/H, $\mathcal{M}_{scal}^{(3)}$ will have the general form:

$$\mathcal{M}_{scal}^{(3)} = \frac{G_3}{H^*}, \tag{4.110}$$

where now, as opposed to the $D = 4$ case, H^* is not the maximal compact subgroup H_3 of the semisimple Lie group G_3, but rather a non-compact real form of the comlexification $H_3^{\mathbb{C}}$ of H_3. This is related to the indefinite signature of the metric on this manifold. The real form H^* is characterized by the property that the symmetric coset H^*/H_c, H_c being the maximal compact subgroup of H^*, has dimension $2n_V$. The isometry group G_3 of $\mathcal{M}_{scal}^{(3)}$ is the global symmetry group of the $D = 3$ theory and is an *off-shell* symmetry since there is no issue about electric-magnetic duality as there was in four dimensions.

Spherically symmetric black-hole solutions are described by *geodesics* $\Phi^I(\tau)$ on $\mathcal{M}_{scal}^{(3)}$ parametrized by the radial variable τ [13]. We shall restrict ourselves to spherically symmetric solutions with vanishing NUT charge, namely we shall take $\omega_\tau = 0$. These solutions will be described by the following effective Lagrangian:

$$\mathcal{L}_3 = \dot{U}^2 + \frac{1}{2}\, G_{rs}\, \dot{\phi}^r\, \dot{\phi}^s + \frac{e^{-2U}}{2}\, \dot{Z}^T\, \mathcal{M}\, \dot{Z} = \frac{1}{2}\, G_{\alpha\beta}\, \dot{q}^\alpha\, \dot{q}^\beta.$$

Let us introduce the following generalized coordinates $q^\alpha \equiv (U, \phi^r, Z^M) = (q^i, Z^M)$. The conjugate momenta p_α will read:

$$p_\alpha = G_{\alpha\beta}(q^i)\, \dot{q}^\alpha = (p_i, p_M). \tag{4.111}$$

In terms of q^α, p_α we write the Hamiltonian:

$$\mathcal{H}_3(p,q) = \frac{1}{2}\, G^{\alpha\beta}\, p_\alpha\, p_\beta = \frac{1}{2}\, G^{ij}(q^i)\, p_i\, p_j + \frac{e^{2U}}{2}\, p_M\, \mathcal{M}^{MN}(\phi^r)\, p_N = c^2,$$

where $\mathcal{M}^{MN}$ denote the entries of the inverse matrix $\mathcal{M}^{-1}$ of $\mathcal{M}_{MN}$, given by: $\mathcal{M}^{-1} = -\mathbb{C}\,\mathcal{M}\,\mathbb{C}$. Since $\mathcal{Z}^M$ are cyclic, the corresponding momenta p_M are constants of motion. They are identified with the quantized charges as follows: $p_M = \mathbb{C}_{MN}\,\Gamma^N$. With this identification, the last term in $\mathcal{H}_3$ reads $\frac{1}{2}\,e^{2U}\,\Gamma^T \cdot \mathcal{M} \cdot \Gamma = -V(q^i, \Gamma)$ and $\mathcal{H}_3$ coincides with the Hamiltonian $\mathcal{H}$ defined in the previous sections. Therefore the resulting equations of motion for p_i, q^i are the same as those discussed earlier. As far as the equation for $\mathcal{Z}^M$ is concerned, it reads [4, 13]:

$$\dot{\mathcal{Z}}^M = \mathbb{F}^M_{t\tau} = \frac{\partial \mathcal{H}_3}{\partial p_M} = e^{2U}\,\mathbb{C}^{MN}\,\mathcal{M}_{NP}\,\Gamma^P . \qquad (4.112)$$

This analysis can be viewed as an extension of that given in [33] since it includes in the definition of the phase space also the four dimensional scalar fields.

In this enlarged Hamiltonian system we want now to define Hamilton's characteristic function, to be denoted by $\mathcal{W}_3(q^\alpha, P_\alpha)$. By definition this function generates the canonical transformation to the coordinates Q^α, P_α, where P_α are constants of motion. Since also c^2 is conserved, it will be a function of the P_α, $c^2 = c^2(P)$. It will indeed provide the Hamiltonian in the new coordinates. From the general theory it is known that the coordinates Q^α are linear in τ, i.e. harmonic functions:

$$Q^\alpha = \left(\frac{\partial c^2}{\partial P_\alpha} \right) \tau + Q^\alpha_0 . \qquad (4.113)$$

If we choose one of the P_i to coincide with c^2, then only the corresponding Q^i will be linear in τ, the other Q^α being constants. The function $\mathcal{W}_3(q^\alpha, P_\alpha)$ satisfies the following relations:

$$p_\alpha = \frac{\partial \mathcal{W}_3}{\partial q^\alpha} , \qquad (4.114)$$

$$Q^\alpha = \frac{\partial \mathcal{W}_3}{\partial P_\alpha} , \qquad (4.115)$$

$$c^2 = \mathcal{H}_3(q^\alpha, \frac{\partial \mathcal{W}_3}{\partial q^\alpha}) , \qquad (4.116)$$

the latter being the Hamilton–Jacobi equation. Since p_M are already constant, $\mathcal{W}_3$ should be such that $P_M = p_M$. This function can be expressed in terms of the four-dimensional Hamilton's characteristic function $\mathcal{W}$ as follows:

$$\mathcal{W}_3(q^\alpha, P_\alpha) = \mathcal{W}(q^i, P_i, P_M) + \mathcal{Z}^M\,P_M , \qquad (4.117)$$

where $\mathcal{W}$ was defined in (4.53). Equation 4.114, for $\alpha = i$, follows from (4.52) and, for $\alpha = M$ implies $p_M = P_M$. Therefore the dependence of $\mathcal{W}$ on P_M is nothing but the dependence of the four-dimensional $\mathcal{W}$ on the quantized charges

$$\Gamma^M = -\mathbb{C}^{MN}\,p_N = -\mathbb{C}^{MN}\,P_N . \qquad (4.118)$$

Equation (4.117) can then be rewritten in the form:

$$\mathcal{W}_3(q^\alpha, P_\alpha) = \mathcal{W}(q^i, P_i, \Gamma^M) + \mathcal{Z}^M \, \mathbb{C}_{MN} \, \Gamma^N . \tag{4.119}$$

Let us now consider the component $\alpha = M$ of Eq. (4.115):

$$\frac{\partial \mathcal{W}_3}{\partial P_M} = \frac{\partial \mathcal{W}}{\partial P_M} + \mathcal{Z}^M = Q^M . \tag{4.120}$$

The above equation can also be written, using (4.118):

$$\mathcal{Z}^M + \mathbb{C}^{MN} \frac{\partial \mathcal{W}}{\partial \Gamma^N} = Q^M . \tag{4.121}$$

This is a non-trivial equation which implies that the combination on the left hand side is a symplectic vector of harmonic functions. Since the Q^M can be chosen to be constant, we can write:

$$\mathcal{Z}^M = -\mathbb{C}^{MN} \frac{\partial \mathcal{W}}{\partial \Gamma^N} + const . \tag{4.122}$$

The above equation allows to compute the electric-magnetic potentials once the $\mathcal{W}$- prepotential is known on the solution as a function of all the quantized charges.

Below we shall illustrate the validity of Eq. (4.122) on the *BPS* solutions in $N = 2$ supergravity.

4.5.1   The BPS Solution for the N = 2 Case

We shall refer to the usual $N = 2$ special geometry notations. Let z^i denote the complex scalar fields on the special Kähler manifold and let $V^M(z, \bar{z})$, $U_i^M(z, \bar{z})$ be the covariantly holomorphic symplectic section and its covariant derivative:

$$V^M = \begin{pmatrix} L^\Lambda \\ M_\Lambda \end{pmatrix} , \quad U_i^M = D_i V^M = \begin{pmatrix} f_i^\Lambda \\ h_{\Lambda i} \end{pmatrix} . \tag{4.123}$$

The matrix $\mathcal{M} = (\mathcal{M}_{MN})$ is related to the above quantities as follows:

$$\mathbb{C} \mathcal{M} \mathbb{C} = -\mathcal{M}^{-1} = V \bar{V}^T + \bar{V} V^T + g^{i\bar{j}} U_i \bar{U}_{\bar{j}}^T + g^{\bar{j}i} \bar{U}_{\bar{j}} U_i^T . \tag{4.124}$$

The symplectic section V^M also satisfies the property: $V^T \mathbb{C} V = i$.
   The central charge Z is defined as follows:

$$Z = -\Gamma^T \mathbb{C} V = e_\Lambda L^\Lambda - m^\Lambda M_\Lambda , \tag{4.125}$$

The first order equations describing the spatial evolution of the BPS solution originate from the Killing-spinor equations and read:

$$\dot{U} = e^U |Z| \, , \quad \dot{z}^i = 2 e^U \, g^{i\bar{j}} \, \partial_{\bar{j}} |Z| \, . \tag{4.126}$$

The corresponding prepotential $\mathcal{W}$ has the following form $\mathcal{W} = 2\, e^U \, |Z|$.

We wish now to verify Eq. (4.122) for this class of solutions. To this end we show that the derivative of the right hand side of this equation equals $\dot{z}$, as given from Eq. (4.112):

$$\frac{d}{d\tau} \frac{\partial \mathcal{W}}{\partial \Gamma^M} = \mathbb{C}_{MN} \, \dot{z}^N = -e^{2U} \, \mathcal{M}_{MN} \, \Gamma^N \, . \tag{4.127}$$

Let us define the quantity:

$$T = -H^T \, \mathbb{C} \, V = H_\Lambda \, L^\Lambda - H^\Lambda \, M_\Lambda \, , \tag{4.128}$$

where we have introduced the symplectic vector H^M of harmonic functions $H^M(\tau) \equiv h^M - \sqrt{2} \, \Gamma^M \, \tau$. In terms of the above quantities, it was shown in [9] that the BPS solution is defined by the following algebraic equations:

$$\bar{T} \, V^M - T \, \bar{V}^M = -\frac{i}{\sqrt{2}} \, H^M \, , \quad e^{-U} = |T| \, , \tag{4.129}$$

with the condition that $H^T \, \mathbb{C} \, \dot{H} = 0$. From the above relations and positions one can prove the following properties:

$$\mathrm{Im}(T \, \bar{Z}) = 0 \, , \quad \dot{T} = -Z \, . \tag{4.130}$$

Differentiating $\mathcal{W} = 2\, e^U \, |Z|$ with respect to Γ and using (4.122) we find

$$z^M = -2 \, \frac{e^U}{|Z|} \, \mathrm{Re}(\bar{Z} \, V^M) = -2 \, e^{2U} \, \mathrm{Re}(\bar{T} \, V^M) \, . \tag{4.131}$$

Using (4.126) and (4.130) one finds:

$$\frac{d}{d\tau} (\bar{T} \, V^M) = -\left(V^M \, \bar{V}^N - g^{i\bar{j}} \, U_i^M \, \bar{U}_{\bar{j}}^N \right) \mathbb{C}_{NP} \, \Gamma^P \, , \tag{4.132}$$

and then

$$\dot{z}^M = -4 \, |Z| \, e^{3U} \, \mathrm{Re}(\bar{T} \, V^M) + e^{2U} \, (V^M \, \bar{V}^N + \bar{V}^M \, V^N - g^{i\bar{j}} \, U_i^M \, \bar{U}_{\bar{j}}^N \\ - g^{\bar{j}i} \, \bar{U}_{\bar{j}}^M \, U_i^N) \, \mathbb{C}_{NP} \, \Gamma^P \, . \tag{4.133}$$

Using (4.129), the first term on the right hand side of the above formula can be rewritten as follows

$$4\,|Z|\,e^{3U}\,\mathrm{Re}(\bar{T}\,V^M) = 2\,e^{2U}\,(V^M\,\bar{V}^N + \bar{V}^M\,V^N)\,\mathbb{C}_{NP}\,\Gamma^P\,. \qquad (4.134)$$

Finally, using (4.124) and the above property, Eq. (4.133) then yields Eq. (4.127).

4.5.2 Geodesics and Duality

Let us end this report by recalling some basic facts about geodesics corresponding to black holes in extended, symmetric supergravities, and their properties with respect to the $D = 3$ global symmetry group (duality group) G_3 [25].

Let us denote by $\mathfrak{g}_3$ and $\mathfrak{H}^*$ the Lie algebras generating G_3 and H^*, respectively. On $\mathfrak{g}_3$ we can define a (pseudo-) Cartan involution σ with respect to which the algebra decompose as follows:

$$\mathfrak{g}_3 = \mathfrak{H}^* \oplus \mathfrak{K}, \qquad (4.135)$$

where the subalgebra $\mathfrak{H}^*$ and its orthogonal complement $\mathfrak{K}$ are defined as the $+1$ and -1 eigenspaces of σ, respectively. Working in a real representation of G_3 one can define an H^*-invariant metric η such that the action of σ on a generator $X \in \mathfrak{g}_3$ reads: $\sigma(X) = -\eta X^T \eta$.

Next we choose a parametrization of our coset. It turns out that the scalar fields Φ^I introduced in the previous subsection are the parameters of a solvable subalgebra $Solv_3$ of $\mathfrak{g}_3$. Such subalgebra is defined by the Iwasawa decomposition of $\mathfrak{g}_3$ with respect to its maximal compact subalgebra $\mathfrak{H}_3$:

$$\mathfrak{g}_3 = \mathfrak{H}_3 \oplus Solv_3\,. \qquad (4.136)$$

Since not every element of G_3 can be written as the product of an element of e^{Solv_3} times an element of H^*, the solvable parametrization in terms of Φ^I is not global on G_3/H^*. This is to be contrasted with what happens for the scalar manifold in four dimensions, since now G_3/H^* is *not* simply connected. The fact that Φ^I are the physical scalars, however, suggests that we should refine our definition of scalar manifold and restrict it to the solvable patch only:

$$\mathcal{M}^{(3)}_{scal} \equiv e^{Solv_3} \subset \frac{G_3}{H^*}\,. \qquad (4.137)$$

Note that e^{Solv_3} is not *geodesically complete* within $\frac{G_3}{H^*}$: There are geodesics on the latter which hit the boundary of the solvable patch. This is where singularities of the corresponding $D = 4$ solution arise: $e^{-U} \to 0$. Therefore if we restrict ourselves to geodesics originating from regular (or small) black holes, they completely unfold in e^{Solv_3}.

Take now a coset representative $\mathcal{V}(\Phi)$, $\Phi \equiv (\Phi^I)$, in e^{Solv_3} and compute it along a geodesic described by the functions $\Phi^I(\tau)$:

$$\mathcal{V}(\tau) \equiv \mathcal{V}(\Phi(\tau)). \tag{4.138}$$

We can evaluate the *pull-back* along the geodesic of the left-invariant one-form $\mathcal{V}^{-1}d\mathcal{V}$ and decompose it in the two subspaces $\mathfrak{H}^*$, $\mathfrak{K}$ of $\mathfrak{g}_3$:

$$\mathcal{V}^{-1}\frac{d}{d\tau}\mathcal{V} = V(\tau) + W(\tau). \tag{4.139}$$

The matrix $V(\tau)$, defined as the pull-back of the vielbein on the geodesic, is called *Lax operator* since it satisfies a Lax-pair equation of the form:

$$\frac{d}{d\tau}V(\tau) = [W(\tau), V(\tau)]. \tag{4.140}$$

This is the equation describing the geodesics on our scalar manifold and $V(\tau)$ can be thought of as the "velocity vector" along the geodesic, so that a geodesic is uniquely defined by a complete set of "initial data", consisting of an "initial point" $\Phi_0 \equiv (\Phi^I(\tau=0))$ and an "initial velocity" $V_0 \equiv V(\tau=0)$, see Fig. 4.2a. The conserved charges of the solution are described by the Noether charge matrix:

$$Q = \mathcal{V}(\tau)V(\tau)\mathcal{V}(\tau)^{-1} = const. \in \mathfrak{g}_3. \tag{4.141}$$

The square of the extremality parameter is related to the Lax matrix as follows:

$$c^2 = \frac{1}{2}G_{IJ}(\Phi)\dot{\Phi}^I\dot{\Phi}^J \propto \text{Tr}(V^2). \tag{4.142}$$

Equation 4.140 suggests that the geodesics are described by a *Liouville integrable system*. This is the case and this property was exploited in the study of black holes in [17], where the complete system of prime integrals in involution was explicitly computed.

The action of the global symmetry group G_3 on a geodesic can be described as follows: By means of a transformation $G_3/H^\star$ we can move the "initial point" Φ_0 at $\tau = 0$ anywhere on the manifold, while for a fixed initial point we can act by means of $H^\star$ on the "initial velocity vector", namely on V_0, see Fig. 4.2b. Since the action of $G_3/H^\star$ is transitive on the scalar manifold, we can always bring the initial point to coincide with the origin (where all the scalar fields vanish) and classify the geodesics according to the $H^\star$-orbit of the Lax matrix at radial infinity V_0. Since the evolution of the Lax operator occurs via a similarity transformation of V_0 by means of a τ-evolving element of the subgroup $H^\star$, it will unfold within a same $H^\star$-orbit. Therefore studying the properties of geodesics with respect to the duality group G_3 amounts to classifying all possible solutions by means *of $H^\star$-orbits within $\mathfrak{K}$.*

Fig. 4.2 Action of G_3 on a geodesic

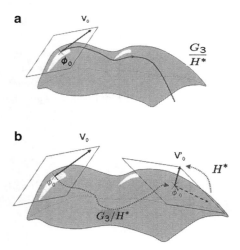

The regularity condition imposes severe constraints on the $H^\star$-orbit, see [25]. In the case of extremal solutions Eq. (4.142) implies $c^2 = 0 = \mathrm{Tr}(V^2)$, and regularity further requires V to be step-k nilpotent, namely

$$V^k = 0 \,, \quad V^{k-1} \neq 0 \,, \tag{4.143}$$

where k should not exceed 5 if V is computed in the adjoint representation of G_3 [25, 28]. The study of extremal solutions and of their properties with respect to the action of G_3 therefore amounts to classifying nilpotent $H^\star$-orbits within $\mathfrak{K}$. This is still an open mathematical problem. Recent progress in this direction was made in [22, 34].

We shall not elaborate further on this issue since it would lead far from main topic of the present dissertation.

4.6   Conclusions

In this report we concentrated only on few issues related to black holes in extended supergravities, trying however to give a self-contained exposition. In particular we discussed a general mathematical characterization of the W-superpotential which defines the "gradient-flow" equations governing the radial evolution of the scalar fields for the extremal static solutions. Such function, besides allowing a specific mathematical description of duality invariant classes of extremal solutions, is likely to have a deeper physical meaning, in relation to their microscopic description. It is duality-invariant and monotonic in τ along the solutions, so that, in the context of QFT/gravity correspondence, it is a candidate for a c-function characterizing the renormalization group flow towards the conformal fixed point [19, 23, 30, 37] (corresponding to the $AdS_2 \times S_2$ near-horizon geometry).

It would be interesting to extend this first-order formalism to rotating or multi-center solutions [8] that is to associate with them a duality invariant W-function (see for instance [26] for a recent work in this direction).

An other interesting direction for further investigation is the first order description of non-extremal static solutions ($c^2 > 0$), whose W-function must have the general form (4.55) (see [4] for comment on this case and for [35] for a recent work).

References

1. D.V. Alekseevsky, Classification of quaternionic spaces with a transitive solvable group of motions. Math. USSR Izv. **9**, 297–339 (1975); B. de Wit, F. Vanderseypen, A. Van Proeyen, Symmetry structure of special geometries. Nucl. Phys. B **400**, 463 (1993) (arXiv:hep-th/9210068); V. Cortés, Alekseevskian spaces. Differ. Geom. Appl. **6**, 129–168 (1996); R. D'Auria, S. Ferrara, M. Trigiante, Critical points of the black-hole potential for homogeneous special geometries. J. High Energy Phys. **0703**, 097 (2007) (arXiv:hep-th/0701090)
2. L. Andrianopoli, M. Bertolini, A. Ceresole, R. D'Auria, S. Ferrara, P. Fre, T. Magri, N = 2 supergravity and N = 2 SuperYang-Mills theory on general scalar manifolds: Symplectic covariance, gaugings and the momentum map. J. Geom. Phys. **23**, 111–189 (1997) (arXiv:hep-th/9605032 (hep-th))
3. L. Andrianopoli, R. D'Auria, S. Ferrara, P. Fre, M. Trigiante, E(7)(7) duality, BPS black hole evolution and fixed scalars. Nucl. Phys. B **509**, 463 (1998) (arXiv:hep-th/9707087); M. Bertolini, M. Trigiante, Regular BPS black holes: Macroscopic and microscopic description of the generating solution. Nucl. Phys. **B582**, 393–406 (2000) (hep-th/0002191)
4. L. Andrianopoli, R. D'Auria, E. Orazi, M. Trigiante, First order description of black holes in moduli space. J. High Energy Phys. **0711**, 032 (2007) (arXiv:0706.0712 (hep-th))
5. For reviews on black holes in superstring theory see: L. Andrianopoli, R. D'Auria, S. Ferrara and M. Trigiante, Extremal black holes in supergravity. Lect. Notes Phys. **737**, 661 (2008) (arXiv:hep-th/0611345); J.M. Maldacena, Black-holes in string theory, hep-th/9607235; A.W. Peet, TASI lectures on black holes in string theory, hep-th/0008241; B. Pioline, Lectures on black holes, topological strings and quantum attractors. Class. Quantum Gravity **23**, S981 (2006) (arXiv:hep-th/0607227); A. Dabholkar, Black hole entropy and attractors. Class. Quantum Gravity **23**, S957 (2006)
6. L. Andrianopoli, R. D'Auria, E. Orazi, M. Trigiante, First order description of D = 4 static black holes and the Hamilton-Jacobi equation. Nucl. Phys. B **833**, 1 (2010) (arXiv:0905.3938 (hep-th)); L. Andrianopoli, R. D'Auria, S. Ferrara, M. Trigiante, Fake superpotential for large and small extremal black holes. J. High Energy Phys. **1008**, 126 (2010) (arXiv:1002.4340 (hep-th))
7. V.I. Arnold, *Mathematical Methods of Classical Mechanics*. Graduate Texts in Mathematics (Springer, New York, 1997); K. Meyer, G. Hall, *Introduction to Hamiltonian Dynamical Systems and the N-Body Problem* (Springer, New York, 1992)
8. D. Astefanesei, K. Goldstein, R.P. Jena, A. Sen, S.P. Trivedi, Rotating attractors. J. High Energy Phys. **0610**, 058 (2006) (arXiv:hep-th/0606244); K. Goldstein, S. Katmadas, Almost BPS black holes. J. High Energy Phys. **0905**, 058 (2009) (arXiv:0812.4183 (hep-th)); I. Bena, G. Dall'Agata, S. Giusto, C. Ruef, N.P. Warner, Non-BPS black rings and black holes in Taub-NUT. J. High Energy Phys. **0906**, 015 (2009). (arXiv:0902.4526 (hep-th)); G. Dall'Agata, S. Giusto, C. Ruef, "U-duality and non-BPS solutions. J. High Energy Phys. **1102**, 074 (2011) (arXiv:1012.4803 (hep-th); L. Andrianopoli, R. D'Auria, S. Ferrara, A. Marrani, M. Trigiante, Two-centered magical charge orbits. J. High Energy Phys. **1104**, 041 (2011) (arXiv:1101.3496

(hep-th)); A. Ceresole, S. Ferrara, A. Marrani, A. Yeranyan, Small black hole constituents and horizontal symmetry. J. High Energy Phys. **1106**, 078 (2011) (arXiv:1104.4652 (hep-th))

9. K. Behrndt, G. Lopes Cardoso, B. de Wit, D. Lust, T. Mohaupt, W.A. Sabra, Higher-order black-hole solutions in $N = 2$ supergravity and Calabi-Yau string backgrounds. Phys. Lett. B **429**, 289 (1998) (arXiv:hep-th/9801081)

10. S. Bellucci, S. Ferrara, M. Gunaydin, A. Marrani, Charge orbits of symmetric special geometries and attractors. Int. J. Mod. Phys. A **21**, 5043 (2006) (arXiv:hep-th/0606209); S. Bellucci, S. Ferrara, R. Kallosh, A. Marrani, Extremal black hole and flux vacua attractors. Lect. Notes Phys. **755**, 115 (2008) (arXiv:0711.4547 (hep-th))

11. L. Borsten, D. Dahanayake, M.J. Duff, W. Rubens, Black holes admitting a freudenthal dual. Phys. Rev. D **80**, 026003 (2009) (arXiv:0903.5517 (hep-th)); L. Borsten, D. Dahanayake, M.J. Duff, S. Ferrara, A. Marrani, W. Rubens, Observations on integral and continuous U-duality orbits in $N = 8$ supergravity. Class. Quantum Gravity **27**, 185003 (2010) (arXiv:1002.4223 (hep-th))

12. G. Bossard, Y. Michel, B. Pioline, Extremal black holes, nilpotent orbits and the true fake superpotential. J. High Energy Phys. **1001**, 038 (2010)

13. P. Breitenlohner, D. Maison, G.W. Gibbons, Four-dimensional black holes from Kaluza-Klein theories. Commun. Math. Phys. **120**, 295 (1988)

14. B.L. Cerchiai, S. Ferrara, A. Marrani, B. Zumino, Duality, entropy and ADM mass in supergravity. Phys. Rev. D **79**, 125010 (2009) (arXiv:0902.3973 (hep-th))

15. A. Ceresole, G. Dall'Agata, Flow equations for non-BPS extremal black holes. J. High Energy Phys. **0703**, 110 (2007) (arXiv:hep-th/0702088); G. Lopes Cardoso, A. Ceresole, G. Dall'Agata, J.M. Oberreuter, J. Perz, First-order flow equations for extremal black holes in very special geometry. J. High Energy Phys. **0710**, 063 (2007) (arXiv:0706.3373 (hep-th))

16. A. Ceresole, G. Dall'Agata, S. Ferrara, A. Yeranyan, First order flows for $N = 2$ extremal black holes and duality invariants. Nucl. Phys. **B824** , 239–253 (2010) (arXiv:0908.1110 (hep-th)); A. Ceresole, G. Dall'Agata, S. Ferrara, A. Yeranyan, Universality of the superpotential for $d = 4$ extremal black holes. Nucl. Phys. **B832**, 358–381 (2010) (arXiv:0910.2697 (hep-th))

17. W. Chemissany, P. Fre, J. Rosseel, A.S. Sorin, M. Trigiante, T. Van Riet, Black holes in supergravity and integrability. J. High Energy Phys. **1009**, 080 (2010) (arXiv:1007.3209 (hep-th)); W. Chemissany, J. Rosseel, T. Van Riet, Black holes as generalised toda molecules. Nucl. Phys. B **843**, 413 (2011) (arXiv:1009.1487 (hep-th))

18. A. Dabholkar, A. Sen, S.P. Trivedi, Black hole microstates and attractor without super-symmetry. J. High Energy Phys. **0701**, 096 (2007) (arXiv:hep-th/0611143); P.K. Tripathy, S.P. Trivedi, Non-supersymmetric attractors in string theory. J. High Energy Phys. **0603**, 022 (2006) (arXiv:hep-th/0511117)

19. J. de Boer, E.P. Verlinde, H.L. Verlinde, On the holographic renormalization group. J. High Energy Phys. **0008**, 003 (2000) (arXiv:hep-th/9912012)

20. S. Ferrara, M. Gunaydin, Orbits of exceptional groups, duality and BPS states in string theory. Int. J. Mod. Phys. A **13**, 2075 (1998) (arXiv:hep-th/9708025)

21. S. Ferrara, R. Kallosh, A. Strominger, N=2 extremal black holes. Phys. Rev. D **52**, 5412 (1995) (arXiv:hep-th/9508072); S. Ferrara, R. Kallosh, Supersymmetry and attractors. Phys. Rev. D **54**, 1514 (1996) (arXiv:hep-th/9602136); S. Ferrara, R. Kallosh, Universality of supersymmetric attractors. Phys. Rev. D **54**, 1525 (1996) (arXiv:hep-th/9603090); S. Ferrara, G.W. Gibbons, R. Kallosh, Black holes and critical points in moduli space. Nucl. Phys. B **500**, 75 (1997) (arXiv:hep-th/9702103); K. Goldstein, N. Iizuka, R.P. Jena, S.P. Trivedi, Non-supersymmetric attractors. Phys. Rev. D **72**, 124021 (2005) (arXiv:hep-th/0507096); R. Kallosh, New attractors. J. High Energy Phys. **0512**, 022 (2005) (arXiv:hep-th/0510024);

22. P. Fre, A.S. Sorin, M. Trigiante, Integrability of supergravity black holes and new tensor classifiers of regular and nilpotent orbits. J. High Energy Phys. **1204**, 015 (2012) arXiv:1103.0848 (hep-th)]; P. Fre, A.S. Sorin, M. Trigiante, Black hole nilpotent orbits and tits satake universality classes. arXiv:1107.5986 ([hep-th)

23. M. Fukuma, S. Matsuura, T. Sakai, Holographic renormalization group. Prog. Theor. Phys. **109**, 489 (2003). (arXiv:hep-th/0212314)

24. M.K. Gaillard, B. Zumino, Duality rotations for interacting fields. Nucl. Phys. B **193**, 221 (1981)
25. D. Gaiotto, W.W. Li, M. Padi, Non-supersymmetric attractor flow in symmetric spaces. J. High Energy Phys. **12**, 093 (2007); E. Bergshoeff, W. Chemissany, A. Ploegh, M. Trigiante, T. Van Riet, Generating geodesic flows and supergravity solutions. Nucl. Phys. **B812**, 343–401 (2009). (arXiv:0806.2310 (hep-th)); G. Bossard, H. Nicolai, K.S. Stelle, Universal BPS structure of stationary supergravity solutions. J. High Energy Phys. **0907**, 003 (2009)
26. P. Galli, K. Goldstein, S. Katmadas, J. Perz, First-order flows and stabilisation equations for non-BPS extremal black holes. J. High Energy Phys. **1106**, 070 (2011) (arXiv:1012.4020 (hep-th))
27. E.G. Gimon, F. Larsen, J. Simon, Black holes in Supergravity: The non-BPS branch. J. High Energy Phys. **0801**, 040 (2008) (arXiv:0710.4967 (hep-th)); G. Lopes Cardoso, A. Ceresole, G. Dall'Agata, J.M. Oberreuter, J. Perz, First-order flow equations for extremal black holes in very special geometry. J. High Energy Phys. **0710**, 063 (2007) (arXiv:0706.3373 (hep-th))
28. M. Gunaydin, A. Neitzke, B. Pioline, A. Waldron, BPS black holes, quantum attractor flows and automorphic forms. Phys. Rev. **D73**, 084019 (2006); B. Pioline, Lectures on black holes, topological strings and quantum attractors, Class. Quantum Gravity **23**, S981 (2006) (hep-th/0607227); M. Gunaydin, A. Neitzke, B. Pioline, A. Waldron, Quantum attractor flows. J. High Energy Phys. **0709**, 056 (2007)
29. W. Hahn, *Stability of Motion* (Springer, Berlin/New York, 1967); N. Rouche, J. Mawhin, *Ordinary Differential Equations*. Stability and Periodic Solutions (Pitman, Boston/London/Melbourne, 1980)
30. K. Hotta, Holographic RG flow dual to attractor flow in extremal black holes. Phys. Rev. **D79**, 104018 (2009) arXiv:0902.3529 (hep-th); (arXiv:0908.1742 (hep-th))
31. C.M. Hull, P.K. Townsend, Unity of superstring dualities. Nucl. Phys. B **438**, 109 (1995). (arXiv:hep-th/9410167).
32. B. Janssen, P. Smyth, T. Van Riet, B. Vercnocke, A first-order formalism for timelike and spacelike brane solutions. J. High Energy Phys. **0804**, 007 (2008). (arXiv:0712.2808 (hep-th))
33. R. Kallosh, From BPS to non-BPS black holes canonically. unpublished arXiv:hep-th/0603003
34. S. Kim, J. Lindman Hörnlund, J. Palmkvist, A. Virmani, Extremal solutions of the S^3 model and nilpotent $G_{2(2)}$ orbits. J. High Energy Phys. **1008**, 072 (2010) (arXiv::1004.5242v2)
35. P. Meessen, T. Ortin, Non-extremal black holes of N = 2, d = 5 supergravity. Phys. Lett. **B707** 178–183 (2012) arXiv:1107.5454 (hep-th)
36. A. Strominger, C. Vafa, Microscopic origin of the Bekenstein-Hawking entropy. Phys. Lett. B **379**, 99 (1996) (arXiv:hep-th/9601029)
37. E.P. Verlinde, H.L. Verlinde, RG-flow, gravity and the cosmological constant. J. High Energy Phys. **0005**, 034 (2000) (arXiv:hep-th/9912018)
38. E. Witten, String theory dynamics in various dimensions. Nucl. Phys. B **443**, 85 (1995) (arXiv:hep-th/9503124)

Chapter 5
On the Classification of Two Center Orbits for Magical Black Holes

Laura Andrianopoli, Riccardo D'Auria, and Sergio Ferrara

We report on recent work [4] concerning the determination of the two-centered generic charge orbits of magical $\mathcal{N} = 2$ and maximal $\mathcal{N} = 8$ supergravity theories in four dimensions. These orbits are classified by seven U-duality invariant polynomials, which group together into four invariants under the horizontal symmetry group $SL(2, \mathbb{R})$. These latter are expected to disentangle different physical properties of the two-centered black-hole system. The invariant with the lowest degree in charges is the symplectic product $\langle Q_1, Q_2 \rangle$, known to control the mutual non-locality of the two centers.

5.1 Introduction

Suprgravity solutions for multi-centered extremal black holes in four dimensions have been recently the subject of a deep investigation. Their study may unravel non-perturbative aspects of string BPS states and their brane interpretation [17]. A generalisation of the *attractor Mechanism* [22, 23] (for a review, see e.g. [2]) has been shown to occur, as firstly pointed out by Denef [16], called split attractor flow for BPS $\mathcal{N} = 2$ black holes [5, 16–18].

L. Andrianopoli (✉) · R. D'Auria
Dipartimento di Fisica, Politecnico di Torino, Corso Duca degli Abruzzi 24, I-10129, Torino, Italy

INFN, Sezione di Torino, Torino, Italy
e-mail: laura.andrianopoli@polito.it; riccardo.dauria@polito.it

S. Ferrara
Physics Department, Theory Unit, CERN, CH-1211, Geneva 23, Switzerland
e-mail: sergio.ferrara@cern.ch

S. Bellucci (ed.), *Supersymmetric Gravity and Black Holes*, Springer Proceedings
in Physics 142, DOI 10.1007/978-3-642-31380-6_5,
© Springer-Verlag Berlin Heidelberg 2013

Attempts to generally classify the two-centered solutions of supergravity theories with symmetric scalar manifolds and electric-magnetic duality (U-duality[1]) symmetry given by classical Lie groups have been considered [13,25–27,34]. In particular, within the framework of the *minimal coupling* [35] of vector multiplets to $\mathcal{N} = 2$ supergravity, it was shown in [25] that different physical properties, such as marginal stability and split attractor flow solutions, can be classified by duality-invariant constraints, which in this case involve two dyonic black-hole charge vectors, and not only one.

This leads one to consider the mathematical issue of the classification of orbits of two (or more) dyonic charge vectors in the context of multi-centered black-hole physics. For the theories treated in [25,26], the charge vector lies in the fundamental representation of $U(1,n)$ (*minimally coupled* $\mathcal{N} = 2$ supergravity [35]) and in the spinor-vector representation of $SL(2,\mathbb{R}) \times SO(q,n)$, corresponding to reducible cubic $\mathcal{N} = 2$ sequence [15,31] for $q = 2$, and to matter-coupled $\mathcal{N} = 4$ supergravity for $q = 6$.

In [25], the two-centered U-invariant polynomials of the *minimally coupled* theory were constructed, and shown to be four (dimension of the adjoint of the two-centered horizontal symmetry $U(2)$). The same was done for the aforementioned cubic sequence in [26], where the number of U-invariants were computed to be seven for $n \geqslant 2$, six for $n = 1$ and five for the irreducible t^3 model.

In the present contribution we report about the investigation, performed in [4], in order to generalize these results to four-dimensional supergravity theories with symmetric *irreducible* scalar manifolds, in particular to the $\mathcal{N} = 8$ maximal theory and to the $\mathcal{N} = 2$ magical models. They share the property that for all these theories the field strengths of the vector fields and their magnetic dual transform in a single (symplectic) irreducible representation of the U-duality group. In this respect, the t^3 model, whose U-duality group is $SL(2,\mathbb{R})$, provides a simple yet interesting example, because it may be obtained both as rank-1 truncation of the *reducible* symmetric models and as first, non-generic element of the sequence of *irreducible* $\mathcal{N} = 2$ symmetric models, which contains the four rank-3 magical supergravity theories mentioned above.

We find that when the stabilizer of a two-centered charge orbit is *non-compact*, the corresponding orbit is *not* unique. As we will consider in Sect. 5.3, this feature is also exhibited by the classification of the orbits of two non-lightlike vectors in a pseudo-Euclidean space $E_{p,q}$ of dimension $p + q$ and signature (p, q). A prominent role is played by an *emergent* horizontal symmetry $SL_h(2,\mathbb{R})$, whose invariants classify all possible two-vector orbits.

The two-centered configurations and the generic (BPS) orbit $\mathcal{O} = SL(2,\mathbb{R})$ of t^3 model were studied in Sect. 7 of [26], in which it was pointed out that, as it occurs also for the one-centered case [7], no stabilizer for the two-centered orbit

---

[1] Here U-duality is referred to as the "continuous" symmetries of [11]. Their discrete versions are the U-duality non-perturbative string theory symmetries introduced by Hull and Townsend [32].

exists.[2] The five components of the spin $s = 2$ horizontal tensor $\mathbf{I}_{abcd}$ (defined in (5.85) below, and explicitly given by (5.88)–(5.92)) form a complete basis of duality-invariant polynomials [26]; as a consequence, the counting (5.2) for $p = 2$-centered black hole solutions in the t^3 model simply reads $5 + 3 - 0 = 4 \times 2$, because $I_{p=2} = 5$ and $\dim_\mathbb{R}(\mathcal{G}_p) = 0$. Moreover, there exist only two independent $[SL_h\,(2, \mathbb{R}) \times SL\,(2, \mathbb{R})]$-invariant polynomials, which can be taken to be the symplectic product $\mathcal{W}$ (of order 2 in charges, defined in (5.82) below) and $\mathbf{I}_6$ (of order 6 in charges, defined in (5.97) below); an alternative choice of basis for the $SL\,(2, \mathbb{R})$-invariant polynomials is thus e.g. given by three components of $\mathbf{I}_{abcd}$ out of the five (5.88)–(5.92), and the two horizontal invariants $\mathcal{W}$ and $\mathbf{I}_6$.

The plan of the paper is as follows.

In Sect. 5.2 we give a group theoretical method (based on progressive branchings of symmetry groups, considered as complex groups) to find the multi-centered charge orbits of a theory with a symmetric scalar manifold; we then apply it to all *irreducible* symmetric cases. The analysis of this section will not depend on the real form of the stabilizer of the orbit, and the results will then hold both for BPS and all the non-BPS orbits of the given model. In Sect. 5.4 we propose a complete basis for U-duality polynomials in the presence of two dyonic black-hole charge vectors in irreducible symmetric models, and we also consider the role of the horizontal symmetry in this framework. Section 5.3 extends the analysis of Sect. 5.2 to different non-compact real forms of the stabilizer of one-centered charge orbits related to Jordan algebras over the octonions, namely to $\mathcal{N} = 8$ theory (whose $\frac{1}{8}$-BPS one-centered stabilizer is $E_{6(2)}$) and for exceptional magical $\mathcal{N} = 2$ theory (whose BPS and non-BPS $\mathcal{J}_4 > 0$ one-centered stabilizers are the compact $E_{6(-78)}$ and the non-compact $E_{6(-14)}$, respectively).

Possible extensions of the present investigation may also cover composite configurations with "small" constituents, as well as a detailed study of the multi-centered charge orbits in $\mathcal{N} = 5, 6$-extended supergravity theories.

5.2 Little Group of p Charge Vectors in Irreducible Symmetric Models

We consider a p-center black hole solution in a Maxwell-Einstein supergravity theory in $d = 4$ space-time dimensions.

The p dyonic black-hole charge vectors can be arranged as

$$\mathbf{Q}_a \equiv \{Q_a^M\}_{M=1,\ldots,f}\,, \tag{5.1}$$

where Q_a^M sits in the irreducible representation $(\mathbf{p}, \mathbf{Sympl}\,(G_4))$ of the group $SL_h\,(p, \mathbb{R}) \times G_4$. $\mathbf{p}$ is the fundamental representation (spanned by the index

---

[2] As it holds for the magical $J_3^\mathbb{R}$ model, see Table I.

$a = 1, \ldots, p)$ of the horizontal symmetry group [26] $SL_h(p, \mathbb{R})$ (see Sect. 5.4), while **Sympl** (G_4) is the symplectic irreducible representation of the black-hole charges, spanned by the index $M = 1, \ldots, f$ of the U-duality group G_4, where $f \equiv \dim_{\mathbb{R}} (\textbf{Sympl}\,(G_4))$.

Suppose there are I_p independent G_4-invariant polynomials constructed out of $\mathbf{Q}_a$, and let $\mathcal{G}_p$ denote the little group of the system of charges, defined as the largest subgroup of G_4 such that $\mathcal{G}_p \, \mathbf{Q}_a = \mathbf{Q}_a \; \forall a$. Then, the following relation[3] holds [26]:

$$I_p + \dim_{\mathbb{R}}(G_4) - \dim_{\mathbb{R}}(\mathcal{G}_p) = f \, p. \qquad (5.2)$$

Some preliminary general observations are in order:

- The group theoretical analysis of the present section does not depend on the real form of G_4 and $\mathcal{G}_p$. We will then generally consider the complex groups. From a physical point of view, the BPS and non-BPS cases in various supergravity theories correspond to different choices of non-compact real forms of $\mathcal{G}_p$ (and of G_4, as well). However, for BPS orbits in $\mathcal{N} = 2$ symmetric models, and in particular for magical models, the stabilizer is always the compact form of the relevant group (see Table 5.1 for the case $p = 2$).
- We shall generally assume $\mathbf{Q}_1$ to be in a representation corresponding to a "large" black hole,[4] namely such that the quartic invariant $\mathcal{I}_4\left(\mathcal{Q}_1^4\right) \neq 0$.
- We shall consider "generic" orbits, in which all I_p invariants are independent.
- There are two relevant cases, corresponding to different behaviors in the counting of invariants:

  (a) The largest subgroup commuting with $\mathcal{G}_p$ inside G_4 is $U(1) \subset G_4$, so that $\mathcal{G}_p \times U(1) \subset G$.
  (b) A $U(1)$ commuting with $\mathcal{G}_p$ inside G_4 does not exist.

  In the case (b), all the singlets in the decomposition of $G_4 \rightarrow \mathcal{G}_p$ correspond to p-center G_4-invariant polynomials of **Sympl** (G_4). On the other hand, in the case (a) the number of $\mathcal{G}_p$-singlets corresponds to the number I_p of p-center G_4-invariant polynomials plus one if some of them are charged with respect to $U(1)$, because one of the singlets can still be acted on by the corresponding $U(1)$-grading.
- The general method for working out $\mathcal{G}_p$ and thus I_p, having solved the problem for $p - 1$ centers, is to consider the pth charge vector $\mathbf{Q}_p$ as transforming in a (reducible) representation of the little group $\mathcal{G}_{p-1}$ of the former $p - 1$ charges, and solve the corresponding one-charge-vector problem.

---

[3]A necessary but not sufficient condition for Eq. (5.2) to hold is $p < f$, such that the p dyonic charge vectors can all be taken to be linearly independent.

[4]Multi-center configurations with "small" constituents [5, 10, 29] can be treated as well, and they will be considered elsewhere.

In the next Subsections we will consider the cases $p = 1$ and $p = 2$ in all irreducible symmetric cases pertaining to supergravity theories in $d = 4$ dimensions (with the exception of the rank-1 t^3 model, treated in [26]). In the case $p = 1$, we will retrieve the well known result $I_{p=1} = 1$, whereas in the $p = 2$ case we will obtain $I_{p=2} = 7$ for all theories under consideration.

Before entering into the details, we just recall that the magic models under consideration are related to Euclidean Jordan algebras $J_3^{\mathbb{A}}$ of degree 3, whose elements are in one-to-one correspondence with the vector fields of $N = 2$ five dimensional theories. Under dimensional reduction to $D = 4$, such a correspondence gets extended to a correspondence between the Freundenthal Triple System $\mathcal{F}(J_3)$ over $J_3^{\mathbb{A}}$ and the four-dimensional vector fields and their duals or, equivalently, the associated electric and magnetic charges. More precisely, one can associate an element of $\mathcal{F}(J_3)$ with electric and magnetic charges $\{q_0, q_i, p^0.p^i\} \in \mathbb{R}^{2n_v+2}$ of an $N = 2$ extremal black hole:

$$\begin{pmatrix} p^0 & p^i \\ q^i & q^0 \end{pmatrix} \in \mathcal{F}(J_3) \tag{5.3}$$

where q_0, p^0 are real numbers while q_i, p^i are the components of Jordan algebra elements along a basis of $J_3^{\mathbb{A}}$.

For the $N = 2$, $D = 4$ theories considered here, the scalar manifolds are associated to the symmetric spaces $E_{7(-25)}/E_6 \times U(1)$, $SO^*(12)/U(6)$, $SU(3,3)/S(U(3) \times U(3))$, $Sp(6)/U(3)$ and are related to Jordan algebras $J_3^{\mathbb{A}}$ over the field $\mathbb{A}$ of octonions, quaternions, complex and real numbers respectively [31]. Remarkably, Table 5.1 is in agreement with the results of [33] (see Table III at page 201 of [33]).

Let us note that all the properties discussed in this section depend on the symplectic representation pertaining to the electric and magnetic charges of the U-duality group (coinciding with the isometry group of the corresponding scalar manifolds), while they do not depend on their real form. This allows to extend the discussion of this section to all (pure) supergravity theories with $N > 4$. In particular, the $N = 2$ model based on $J_3^{\mathbb{C}}$ can be treated together with the $N = 5$ theory, as both theories have charges in the **20** of different real forms of $SL(6, \mathbb{C})$, and the $N = 2$ model based on $J_3^{\mathbb{O}}$ can be treated together with the $N = 8$ theory, both models having charges in the **56** of different real forms of E_7. In a similar way, the $N = 2$ model based on $J_3^{\mathbb{H}}$, which has scalar manifold $SO^*(12)/U(6)$ and 32 electric and magnetic field strengths (lying in the spinor representation of $SO^*(12)$), can be treated at the same time with the $N = 6$ theory, having exactly the same bosonic field content and scalar manifold.[5]

---

[5]In this respect, we observe that the $N = 3$ theory coupled to three vector multiplets, whose scalar manifold, $SU(3, 3)/S(U(3) \times U(3))$, coincides with the one of the magic model based on $J_3^{\mathbb{C}}$, cannot be included in the discussion, since the bosonic field content of the two theories is different.

Table 5.1 BPS generic charge orbits of two-centered extremal black holes in $\mathcal{N} = 2$, $d = 4$ magical models. $Conf\left(J_3^\mathbb{A}\right)$ denotes the "conformal" group of $J_3^\mathbb{A}$ (see e.g. [30], and Refs. therein). By introducing $\mathbb{A} = \mathbb{R}$, $\mathbb{C}$, $\mathbb{H}$, $\mathbb{O}$, it is worth remarking that the stabilizer group $\mathcal{G}_{p=2}\left(J_3^\mathbb{A}\right)$ and the automorphism group $Aut\left(\mathbf{t}\left(\mathbb{A}\right)\right)$ of the *normed triality* $\mathbf{t}\left(\mathbb{A}\right)$ in dimension $\dim_\mathbb{R}\mathbb{A} = 1$, 2, 4, 8 (given e.g. in Eq. (5) of [6]) share the same Lie algebra. In other words, $\mathfrak{g}_{p=2}\left(J_3^\mathbb{A}\right) \sim$ tri $(\mathbb{A})$, where tri $(\mathbb{A})$ denotes the Lie algebra of $Aut\left(\mathbf{t}\left(\mathbb{A}\right)\right)$ itself (see e.g. Eq. (21) of [6])

$J_3^\mathbb{A}$	$\mathcal{O}_{p=2,BPS} = \dfrac{Conf\left(J_3^\mathbb{A}\right)}{\mathcal{G}_{p=2}\left(J_3^\mathbb{A}\right)}$
$J_3^\mathbb{O}$	$\dfrac{E_{7(-25)}}{SO(8)}$
$J_3^\mathbb{H}$	$\dfrac{SO^*(12)}{[SU(2)]^3}$
$J_3^\mathbb{C}$	$\dfrac{SU(3,3)}{[U(1)]^2}$
$J_3^\mathbb{R}$	$Sp\left(6, \mathbb{R}\right)$

5.2.1 $J_3^\mathbb{O}$ $(\mathcal{N} = 2)$, $J_3^{\mathbb{O}_s}$ $(\mathcal{N} = 8)$

Let us start considering the exceptional case, based on the Euclidean degree-3 Jordan algebra $J_3^\mathbb{O}$ on the octonions $\mathbb{O}$. Since, as mentioned earlier, we actually work with complex groups, this case pertains also to maximal $\mathcal{N} = 8$ supergravity, based on the Euclidean degree-3 Jordan algebra $J_3^{\mathbb{O}_s}$ on the split octonions $\mathbb{O}_s$.

In the complex field, $G_4 = E_7$ and $\mathbf{Sympl}\left(E_7\right) = \mathbf{Fund}\left(E_7\right) = \mathbf{56}$.

- Let us first solve the one-center problem ($p = 1$). $\mathcal{G}_1$ is a real form of E_6; the **56** branches with respect to E_6 as follows (subscripts denote the U (1)-charges throughout):

$$\mathbf{56} \rightarrow \mathbf{1}_{-3} + \mathbf{27}_{-1} + \overline{\mathbf{27}}_{+1} + \mathbf{1}_{+3}, \tag{5.4}$$

and correspondingly the charge vector $\mathbf{Q}_1$ (defined as $\left(p^\Lambda, q_\Lambda\right)$ throughout) decomposes as follows:

$$\mathbf{Q}_1 = \left(p^0, \mathbf{p}_{27}, q_0, \mathbf{q}_{\overline{27}}\right). \tag{5.5}$$

Note that the branching (5.4) contains two E_6-singlets, and $E_7 \supset E_6 \times U(1) = \mathcal{G}_1 \times U(1)$. According to the previous discussion, one of the singlets can be freely acted on by the $U(1)$. Thus, by acting with $G_4/\mathcal{G}_1 = E_7/E_6$, the one-center charge vector $\mathbf{Q}_1$ can be reduced as follows:

$$\mathbf{Q}_1 \overset{E_7/E_6}{\longrightarrow} (I^{(1)}, \mathbf{0}_{27}, \pm I^{(1)}, \mathbf{0}_{\overline{27}}). \tag{5.6}$$

One is then left with only one independent singlet charge $I^{(1)}$ related to the one-center quartic invariant $\mathcal{I}_4 (\mathcal{Q}_1^4)$; therefore, $I_1 = 1$, as expected. This analysis is consistent with the general formula (5.2), which in this case reads:

$$I_1 + \dim_{\mathbb{R}}(E_7) - \dim_{\mathbb{R}}(E_6) = 1 + 133 - 78 = 56. \tag{5.7}$$

- Let us now proceed to deal with the two charge-vector problem ($p = 2$). The second charge vector is denoted as $\mathbf{Q}_2 \equiv (m^\Lambda, e_\Lambda)$ throughout. Having solved the problem for $p = 1$, we can decompose $\mathbf{Q}_2$ with respect to $\mathcal{G}_1 = E_6$ using (5.4), obtaining the decomposition

$$\mathbf{Q}_2 = (I^{(2)}, \mathbf{m}_{27}, I^{(3)}, \mathbf{e}_{\overline{27}}), \tag{5.8}$$

and then determine the corresponding little group inside E_6. The little group of the irreducible representation $\mathbf{27}$ of E_6 is F_4, under which

$$\mathbf{27} \to \mathbf{1} + \mathbf{26}, \tag{5.9}$$

and correspondingly

$$\mathbf{m}_{27} \to (I^{(4)}, \mathbf{m}_{26}); \quad \mathbf{e}_{\overline{27}} \to (I^{(5)}, \mathbf{e}_{26}). \tag{5.10}$$

Note in particular that F_4 is a maximal (symmetric) subgroup of E_6, so that all singlets correspond to extra E_7-invariant polynomials, and that $\mathbf{m}_{26}$ can be set to zero through the action of $\mathcal{G}_1/F_4 = E_6/F_4$, thus yielding the result:

$$\mathbf{Q}_2 \overset{E_6/F_4}{\longrightarrow} (I^{(2)}, I^{(4)}, \mathbf{0}_{26}, I^{(3)}, I^{(5)}, \mathbf{e}_{26}). \tag{5.11}$$

- The $\mathbf{26}$ of F_4 has little group $SO(8)$, which does not commute with a $U(1)$ in F_4. Under this non-maximal embedding, the $\mathbf{26}$ branches as

$$\mathbf{26} \to \mathbf{1} + \mathbf{1} + \mathbf{8}_v + \mathbf{8}_s + \mathbf{8}_c, \tag{5.12}$$

and correspondingly

$$\mathbf{e}_{26} \to (I^{(6)}, I^{(7)}, \mathbf{e}_{8_v}, \mathbf{e}_{8_s}, \mathbf{e}_{8_c}). \tag{5.13}$$

Therefore, by acting with $F_4/\mathcal{G}_2 = F_4/SO(8)$, $\mathbf{Q}_2$ can then be put in the form

$$\mathbf{Q}_2 \xrightarrow{F_4/SO(8)} (I^{(2)}, I^{(4)}, \mathbf{0}_{26}, I^{(3)}, I^{(5)}, I^{(6)}, I^{(7)}, \mathbf{0}_{8_v}, \mathbf{0}_{8_s}, \mathbf{0}_{8_c}). \tag{5.14}$$

In conclusion, we found that the little group of a two-centered black-hole solution is $\mathcal{G}_2 = SO(8)$, and the corresponding two-centered charge orbits correspond to different real forms of the quotient of complex groups

$$\mathcal{O}_{p=2} = \frac{G_4}{\mathcal{G}_2} = \frac{E_7}{SO(8)}. \tag{5.15}$$

The E_7-invariant polynomials for a two-centered configuration are seven: $I_2 = 7$; indeed, the general formula (5.2) gives:

$$I_2 + \dim_{\mathbb{R}}(E_7) - \dim_{\mathbb{R}}(SO(8)) = 7 + 133 - 28 = 112 = 2 \cdot 56. \tag{5.16}$$

5.2.2 $J_3^{\mathbb{H}}$ ($\mathcal{N} = 2 \leftrightarrow \mathcal{N} = 6$)

This model is based on the Euclidean degree-3 Jordan algebra $J_3^{\mathbb{H}}$ on the quaternions $\mathbb{H}$, and it is "dual" to $\mathcal{N} = 6$ "pure" theory, because these theories share the same bosonic sector [1,3,24,31,37].

In the complex field $G_4 = SO(12)$, and $\mathbf{Sympl}(SO(12)) = \mathbf{32}$, the chiral spinor irreducible representation of $SO(12)$.

- Let us first solve the problem for $p = 1$. $\mathcal{G}_1$ is a real form of $SU(6)$, the relevant (maximal symmetric) embedding is

$$SO(12) \supset SU(6) \times U(1) = \mathcal{G}_1 \times U(1), \tag{5.17}$$

and the $\mathbf{32}$ accordingly branches

$$\mathbf{32} \to \mathbf{1}_{-3} + \mathbf{15}_{-1} + \overline{\mathbf{15}}_{+1} + \mathbf{1}_{+3}, \tag{5.18}$$

corresponding to the charge decomposition

$$\mathbf{Q}_1 = (p^0, \mathbf{p}_{15}, q_0, \mathbf{q}_{15}). \tag{5.19}$$

The analysis here is completely analogous to the exceptional case above. The branching (5.18) contains two $SU(6)$-singlets, but, by virtue of (5.17), one of the singlets can be freely acted on by the $U(1)$. By acting with $G_4/\mathcal{G}_1 = SO(12)/SU(6)$, $\mathbf{Q}_1$ can be reduced to

$$\mathbf{Q}_1 \overset{SO(12)/SU(6)}{\longrightarrow} \left(I^{(1)}, \mathbf{0_{15}}, \pm I^{(1)}, \mathbf{0_{15}}\right), \tag{5.20}$$

so that $I_1 = 1$, corresponding to the one-center quartic invariant $\mathcal{I}_4\left(Q_1^4\right)$ only. Indeed, the general formula (5.2) yields

$$I_1 + \dim_{\mathbb{R}}(SO(12)) - \dim_{\mathbb{R}}(SU(6)) = 1 + 66 - 35 = 32\,. \tag{5.21}$$

- Let us consider now the two-centered case ($p = 2$). Having solved the problem for $p = 1$, we further decompose $\mathbf{Q}_2$ with respect to $\mathcal{G}_1 = SU(6)$:

$$\mathbf{Q}_2 = \left(I^{(2)}, \mathbf{m_{15}}, I^{(3)}, \mathbf{e_{15}}\right), \tag{5.22}$$

and find the corresponding little group. The little group of the **15** of $SU(6)$ is $USp(6)$, under which such a representation branches as follows:

$$\mathbf{15} \longrightarrow \mathbf{1} + \mathbf{14}, \tag{5.23}$$

yielding the charge decompositions

$$\mathbf{m_{15}} \longrightarrow \left(I^{(4)}, \mathbf{m_{14}}\right); \quad \mathbf{e_{15}} \longrightarrow \left(I^{(5)}, \mathbf{e_{14}}\right). \tag{5.24}$$

Since $USp(6)$ is maximally (and symmetrically) embedded in $SU(6)$, all singlets correspond to extra $SO(12)$-invariant polynomials, and $\mathbf{m_{14}}$ can be set to zero through the action of $\mathcal{G}_1/USp(6) = SU(6)/USp(6)$, thus yielding the result:

$$\mathbf{Q}_2 \overset{SU(6)/USp(6)}{\longrightarrow} \left(I^{(2)}, I^{(4)}, \mathbf{0_{14}}, I^{(3)}, I^{(5)}, \mathbf{e_{14}}\right). \tag{5.25}$$

- The **14** (rank-2 antisymmetric) of $USp(6)$ has little group $[SU(2)]^3$, which does not commute with a $U(1)$ in $USp(6)$. The **14** correspondingly branches as

$$\mathbf{14} \longrightarrow (\mathbf{1,1,1}) + (\mathbf{1,1,1}) + (\mathbf{1,2,2}) + (\mathbf{2,2,2})\,, \tag{5.26}$$

and thus

$$\mathbf{e_{14}} \longrightarrow \left(I^{(6)}, I^{(7)}, \mathbf{e_{(1,2,2)}}, \mathbf{e_{(2,2,2)}}\right). \tag{5.27}$$

Therefore, by acting with $USp(6)/\mathcal{G}_2 = USp(6)/[SU(2)]^3$, $\mathbf{Q}_2$ can then be put in the form

$$\mathbf{Q}_2 \overset{USp(6)/[SU(2)]^3}{\longrightarrow} \left(I^{(2)}, I^{(4)}, \mathbf{0_{14}}, I^{(3)}, I^{(5)}, I^{(6)}, I^{(7)}, \mathbf{0_{(1,2,2)}}, \mathbf{0_{(2,2,2)}}\right). \tag{5.28}$$

In conclusion, we found that the little group of a two-centered black-hole solution is $\mathcal{G}_2 = [SU(2)]^3$, and the corresponding two-centered charge orbit reads (in complexified form)

$$\mathcal{O}_{p=2} = \frac{G_4}{\mathcal{G}_2} = \frac{SO\,(12)}{[SU\,(2)]^3}. \tag{5.29}$$

The $SO\,(12)$-invariant polynomials for a two-centered configuration are seven: $I_2 = 7$; indeed, the general formula (5.2) gives:

$$I_2 + \dim_{\mathbb{R}}(SO\,(12)) - \dim_{\mathbb{R}}([SU\,(2)]^3) = 7 + 66 - 9 = 64 = 2 \cdot 32. \tag{5.30}$$

5.2.3 $J_3^{\mathbb{C}}$ $(\mathcal{N} = 2)$, $M_{1,2}$ $(\mathbb{O})$ $(\mathcal{N} = 5)$

Let us now consider the model based on the Euclidean degree-3 Jordan algebra $J_3^{\mathbb{C}}$ on $\mathbb{C}$. Since, as mentioned earlier, we actually deal with groups on the complex field, this case pertains also to "pure" $\mathcal{N} = 5$ supergravity, which is based on $M_{1,2}$ $(\mathbb{O})$, the Jordan triple system (not upliftable to $d = 5$) generated by 2×1 matrices over $\mathbb{O}$ [31].

In the complex field, $G_4 = SU\,(6)$, and **Sympl** $(SU\,(6)) = \mathbf{20}$, the real self-dual rank-3 antisymmetric irreducible representation.

- Let us first solve the problem for $p = 1$. $\mathcal{G}_1$ is a real form of $SU(3) \times SU(3)$, the relevant (maximal symmetric) embedding is

$$SU(6) \supset SU(3) \times SU(3) \times U(1) = \mathcal{G}_1 \times U(1), \tag{5.31}$$

and the **20** accordingly branches as

$$\mathbf{20} \to (\mathbf{1},\mathbf{1})_{-3} + (\mathbf{3},\overline{\mathbf{3}})_{-1} + (\overline{\mathbf{3}},\mathbf{3})_{+1} + (\mathbf{1},\mathbf{1})_{+3}, \tag{5.32}$$

corresponding to the charge decomposition

$$\mathbf{Q}_1 \to (p^0, \mathbf{p}_{(3,\overline{3})}, q_0, \mathbf{q}_{(\overline{3},3)}). \tag{5.33}$$

The analysis here is analogous to the cases treated above. The branching (5.32) contains two $[SU(3) \times SU(3)]$-singlets, but, by virtue of (5.31), one of the singlets can be freely acted on by the $U(1)$. By acting with $G_4/\mathcal{G}_1 = SU\,(6)\,/\,[SU(3) \times SU(3)]$, $\mathbf{Q}_1$ can be reduced to

$$\mathbf{Q}_1 \xrightarrow{SU(6)/[SU(3)\times SU(3)]} (I^{(1)}, \mathbf{0}_{(3,\overline{3})}, \pm I^{(1)}, \mathbf{0}_{(\overline{3},3)}), \tag{5.34}$$

so that $I_1 = 1$, which corresponds to $\mathcal{I}_4\,(\mathcal{Q}_1^4)$ only. Indeed, formula (5.2) yields

$$I_1 + \dim_{\mathbb{R}}(SU\,(6)) - \dim_{\mathbb{R}}(SU(3) \times SU(3)) = 1 + 35 - 16 = 20. \tag{5.35}$$

- Let us consider now the two-centered case ($p = 2$). Having solved the problem for $p = 1$, we further decompose $\mathbf{Q}_2$ with respect to $\mathcal{G}_1 = SU(3) \times SU(3)$:

$$\mathbf{Q}_2 = (I^{(2)}, \mathbf{m}_{(3,\bar{3})}, I^{(3)}, \mathbf{e}_{(\bar{3},3)}), \tag{5.36}$$

and find the corresponding little group. The little group of the $(\mathbf{3}, \bar{\mathbf{3}})$ of $SU(3) \times SU(3)$ is the *diagonal* $SU(3)$, which is maximal in $SU(3) \times SU(3)$ (see e.g. [38]), under which such a representation branches as follows:

$$(\mathbf{3}, \bar{\mathbf{3}}) \rightarrow \mathbf{1} + \mathbf{8}, \tag{5.37}$$

yielding the charge decompositions

$$\mathbf{m}_{(3,\bar{3})} \rightarrow (I^{(4)}, \mathbf{m}_8); \quad \mathbf{e}_{(\bar{3},3)} \rightarrow (I^{(5)}, \mathbf{e}_8).$$

The maximality of the embedding of the diagonal $SU(3)$ in $SU(3) \times SU(3)$ implies all singlets to correspond to extra $SU(6)$-invariant polynomials, and $\mathbf{m}_8$ can be set to zero through the action of $\mathcal{G}_1/SU(3) = [SU(3) \times SU(3)]/SU(3)$, thus yielding the result:

$$\mathbf{Q}_2 \xrightarrow{[SU(3) \times SU(3)]/SU(3)} (I^{(2)}, I^{(4)}, \mathbf{0}_8, I^{(3)}, I^{(5)}, \mathbf{e}_8). \tag{5.38}$$

- The $\mathbf{8}$ (adjoint) of $SU(3)$ has little group $[U(1)]^2$, which does not commute with any $U(1)$ in $SU(3)$. The $\mathbf{8}$ correspondingly branches as

$$\mathbf{8} \rightarrow \mathbf{1}_{0,0} + \mathbf{1}_{0,0} + \mathbf{1}_{0,2} + \mathbf{1}_{0,-2} + \mathbf{1}_{3,1} + \mathbf{1}_{3,-1} + \mathbf{1}_{-3,1} + \mathbf{1}_{-3,-1}, \tag{5.39}$$

and thus

$$\mathbf{e}_8 \longrightarrow (I^{(6)}, I^{(7)}, \mathbf{e}_{0,2}, \mathbf{e}_{0,-2}, \mathbf{e}_{3,1}, \mathbf{e}_{3,-1}, \mathbf{e}_{-3,1}, \mathbf{e}_{-3,-1}). \tag{5.40}$$

Therefore, by acting with $SU(3)/\mathcal{G}_2 = SU(3)/[U(1)]^2$, $\mathbf{Q}_2$ can then be put in the form

$$\mathbf{Q}_2 \xrightarrow{SU(3)/[U(1)]^2} (I^{(2)}, I^{(4)}, \mathbf{0}_8, I^{(3)}, I^{(5)}, I^{(6)}, I^{(7)}, \mathbf{0}_6), \tag{5.41}$$

where $\mathbf{0}_6$ collectively denotes the six charges pertaining to the $[U(1)]^2$-charged representations $\mathbf{1}_{0,2}, \mathbf{1}_{0,-2}, \mathbf{1}_{3,1}, \mathbf{1}_{3,-1}, \mathbf{1}_{-3,1}, \mathbf{1}_{-3,-1}$ in the right-hand side of (5.39).

In conclusion, we found that the little group of a two-centered black-hole solution is $\mathcal{G}_2 = [U(1)]^2$, and the corresponding two-centered charge orbit reads (in complexified form)

$$\mathcal{O}_{p=2} = \frac{G_4}{\mathcal{G}_2} = \frac{SU(6)}{[U(1)]^2}. \tag{5.42}$$

The $SU(6)$-invariant polynomials for a two-centered configuration are seven: $I_2 = 7$; indeed, the general formula (5.2) gives:

$$I_2 + \dim_{\mathbb{R}}(SU(6)) - \dim_{\mathbb{R}}([U(1)]^2) = 7 + 35 - 2 = 40 = 2 \cdot 20. \qquad (5.43)$$

5.2.4 $J_3^{\mathbb{R}} (\mathcal{N} = 2)$

Finally, we consider the model based on the Euclidean degree-3 Jordan algebra $J_3^{\mathbb{R}}$ on $\mathbb{R}$.

In the complex field, $G_4 = USp(6)$, and **Sympl**$(USp(6)) = \mathbf{14'}$, the real rank-3 antisymmetric irreducible representation of $USp(6)$ (not to be confused with the rank-2 antisymmetric irreducible representation $\mathbf{14}$ considered in Sect. 5.2.2).

- Let us first solve the problem for $p = 1$. $\mathcal{G}_1$ is a real form of $SU(3)$, the relevant (maximal symmetric) embedding is

$$USp(6) \supset SU(3) \times U(1) = \mathcal{G}_1 \times U(1), \qquad (5.44)$$

  and the $\mathbf{14'}$ accordingly branches as

$$\mathbf{14'} \rightarrow \mathbf{1}_{-3} + \mathbf{6}_{-1} + \overline{\mathbf{6}}_{+1} + \mathbf{1}_{+3}, \qquad (5.45)$$

  corresponding to the charge decomposition

$$\mathbf{Q}_1 \rightarrow (p^0, \mathbf{p_6}, q_0, \mathbf{q_6}). \qquad (5.46)$$

  Once again, the analysis here is analogous to the cases treated above. The branching (5.45) contains two $SU(3)$-singlets, but, by virtue of (5.44), one of the singlets can be freely acted on by the $U(1)$. By acting with $G_4/\mathcal{G}_1 = USp(6)/SU(3)$, $\mathbf{Q}_1$ can be reduced to

$$\mathbf{Q}_1 \xrightarrow{USp(6)/SU(3)} (I^{(1)}, \mathbf{0_6}, \pm I^{(1)}, \mathbf{0_6}), \qquad (5.47)$$

  so that $I_1 = 1$, which corresponds to $\mathcal{I}_4 (\mathcal{Q}_1^4)$ only. Indeed, formula (5.2) yields

$$I_1 + \dim_{\mathbb{R}}(USp(6)) - \dim_{\mathbb{R}}(SU(3)) = 1 + 21 - 8 = 14. \qquad (5.48)$$

- Let us consider now the two-centered case ($p = 2$). Having solved the problem for $p = 1$, we further decompose $\mathbf{Q}_2$ with respect to $\mathcal{G}_1 = SU(3)$:

$$\mathbf{Q}_2 = (I^{(2)}, \mathbf{m_6}, I^{(3)}, \mathbf{e_6}), \qquad (5.49)$$

and find the corresponding little group. The little group of the **6** of $SU(3)$ is $SO(3)$,which is maximal in $SU(3)$, under which such a representation branches as follows:

$$\mathbf{6} \rightarrow \mathbf{1} + \mathbf{5}, \tag{5.50}$$

yielding the charge decompositions

$$\mathbf{m_6} \rightarrow (I^{(4)}, \mathbf{m_5}); \quad \mathbf{e_6} \rightarrow (I^{(5)}, \mathbf{e_5}). \tag{5.51}$$

The maximality of $SO(3)$ in $SU(3)$ implies all singlets to corresponds to extra USp (6)-invariant polynomials, and $\mathbf{m_6}$ can be set to zero through the action of $\mathcal{G}_1/SO(3) = SU(3)/SO(3)$, thus yielding the result:

$$\mathbf{Q_2} \xrightarrow{SU(3)/SO(3)} (I^{(2)}, I^{(4)}, \mathbf{0_5}, I^{(3)}, I^{(5)}, \mathbf{e_5}). \tag{5.52}$$

- Note, however, that the little group of the **5** (rank-2 symmetric traceless) irreducible representation of $SO(3)$ is the identity, so that $\mathcal{G}_2 = \mathbb{I}$. The **5** then trivially branches into five singlets, three of which can be rotated to zero through the action of $SO(3)/\mathcal{G}_2 = SO(3)$:

$$\mathbf{Q_2} \xrightarrow{SO(3)} (I^{(2)}, I^{(4)}, \mathbf{0_5}, I^{(3)}, I^{(5)}, I^{(6)}, I^{(7)}, \mathbf{0_3}), \tag{5.53}$$

where $\mathbf{0_3}$ collectively denotes such three singlets set to zero.

In conclusion, we found that the little group of a two-centered black-hole solution is the identity itself: $\mathcal{G}_2 = \mathbb{I}$, and the corresponding two-centered charge orbit reads (in compact form)

$$\mathcal{O}_{p=2} = \frac{G_4}{\mathcal{G}_2} = USp\,(6)\,. \tag{5.54}$$

The USp (6)-invariant polynomials for a two-centered configuration are seven: $I_2 = 7$; indeed, the general formula (5.2) yields:

$$I_2 + \dim_{\mathbb{R}}(USp(6)) - \dim_{\mathbb{R}}(\mathbb{I}) = 7 + 21 - 0 = 28 = 2 \cdot 14. \tag{5.55}$$

5.3 Two-Centered Orbits with Non-compact Stabiliser: the $\mathcal{N} = 8$ BPS and Octonionic $\mathcal{N} = 2$ Non-BPS Cases

For $\mathcal{N} = 2$ BPS two-centered extremal black holes, the stabiliser of the supporting charge orbit is always *compact*, so the orbit is unique (see Table 5.1 for magical models). This is no longer the case when the stabiliser is non-compact, as it holds for $\mathcal{N} = 2$ two-centered solutions with two non-BPS centers characterised by $\mathcal{I}_4\left(\mathcal{Q}_1^4\right) > 0$ and $\mathcal{I}_4\left(\mathcal{Q}_2^4\right) > 0$, and for $\mathcal{N} \geqslant 3$ two-centered solutions with two $\frac{1}{N}$-BPS centers.

These are interesting cases, in which a *split attractor flow* through a wall of marginal stability has been shown to occur [9,21].

We will consider here the $\frac{1}{8}$- BPS two-centered orbits in the maximal $\mathcal{N} = 8$ theory (based on $J_3^{O_s}$) and the non-BPS two-centered orbits (of the aforementioned type) in the exceptional $\mathcal{N} = 2$ magic model, based on J_3^{O}. These two cases can be obtained by repeating the analysis of Sect. 5.2.1 and choosing suitable non-compact real forms of G_4 and $\mathcal{G}_2$.

The one-centered charge orbits respectively read [7,20]:

$$\mathcal{N} = 8, \; \frac{1}{8}\text{-BPS} : \mathcal{O}_{p=1} = \frac{E_{7(7)}}{E_{6(2)}}; \tag{5.56}$$

$$\mathcal{N} = 2, \; J_3^{O} \text{ nBPS } \mathcal{J}_4 > 0 : \mathcal{O}_{p=1} = \frac{E_{7(-25)}}{E_{6(-14)}}. \tag{5.57}$$

In the maximal case, the chain of relevant group branchings reads

$$\mathcal{N} = 8, \; \frac{1}{8}\text{-BPS} : E_{7(7)} \longrightarrow E_{6(2)} \longrightarrow F_{4(4)} \longrightarrow SO\,(5,4) \longrightarrow \begin{cases} SO\,(4,4) \\ or \\ SO\,(5,3) \end{cases}, \tag{5.58}$$

such that two $\frac{1}{8}$-BPS, $\mathcal{N} = 8$, two-centered charge orbits exist:

$$\mathcal{O}_{\mathcal{N}=8,\frac{1}{8}\text{-BPS},p=2,\mathbf{I}} = \frac{E_{7(7)}}{SO\,(4,4)} \tag{5.59}$$

$$\mathcal{O}_{\mathcal{N}=8,\frac{1}{8}\text{-BPS},p=2,\mathbf{II}} = \frac{E_{7(7)}}{SO\,(5,3)}. \tag{5.60}$$

In the $\mathcal{N} = 2$ exceptional case, the chain of relevant group branchings reads

$$\mathcal{N} = 2, \; J_3^{O} \text{ nBPS} : E_{7(-25)} \longrightarrow E_{6(-14)} \longrightarrow F_{4(-20)} \longrightarrow \begin{cases} SO\,(9) \longrightarrow SO\,(8) \\ or \\ SO\,(8,1) \longrightarrow \begin{array}{l} SO\,(8) \\ or \\ SO\,(7,1) \end{array} \end{cases}, \tag{5.61}$$

such that two non-BPS, $\mathcal{N} = 2$, two-centered charge orbits exist:

$$\mathcal{O}_{\mathcal{N}=2,J_3^{O},\text{nBPS},p=2,\mathbf{I}} = \frac{E_{7(-25)}}{SO\,(8)} \tag{5.62}$$

$$\mathcal{O}_{\mathcal{N}=2,J_3^{O},\text{nBPS},p=2,\mathbf{II}} = \frac{E_{7(-25)}}{SO\,(7,1)}. \tag{5.63}$$

As it holds for the stabilizer of $\mathcal{O}_{\mathcal{N}=2,J_3^{O},\text{BPS},p=2}$ (see Table 5.1), the Lie algebra $\mathfrak{so}\,(8)$ of the stabilizer of $\mathcal{O}_{\mathcal{N}=2,J_3^{O},\text{nBPS},p=2,\mathbf{I}}$ (5.62) is nothing but the Lie algebra tri ($\mathbb{O}$) of the automorphism group Aut ($\mathbf{t}$ ($\mathbb{O}$)) of the *normed triality* over the octonionic division algebra $\mathbb{O}$ (see e.g. Eq. (21) of [6]). It is worth here to observe that the Lie algebra $\mathfrak{so}$ (4, 4) of the stabilizer of $\mathcal{O}_{N=8,\frac{1}{8}\text{-BPS},p=2,\mathbf{I}}$ (5.59) enjoys an analogous interpretation as the Lie algebra tri ($\mathbb{O}_s$) of the automorphism group Aut ($\mathbf{t}$ ($\mathbb{O}_s$)) of the *normed triality* over the *split* form $\mathbb{O}_s$ of the octonions. On the other hand, a similar interpretation seems not to hold for the stabilizer of $\mathcal{O}_{N=8,\frac{1}{8}\text{-BPS},p=2,\mathbf{II}}$ (5.60) as well as for the stabilizer of $\mathcal{O}_{N=2,J_3^{\mathbb{O}},\text{nBPS},p=2,\mathbf{II}}$ (5.63).

We expect the $N = 8$ orbits (5.59) and (5.60), as well as the $N = 2$ orbits (5.62) and (5.63), to be defined by different constraints on the four SL_h $(2,\mathbb{R}) \times G_4$ invariant polynomials given by Eq. (5.101); we leave this interesting issue for further future investigation.

Here, we confine ourselves to present parallel results on pseudo-orthogonal groups, which may shed some light on the whole framework. Let us consider two vectors $\mathbf{x}$ and $\mathbf{y}$ in a pseudo-Euclidean $(p + q)$-dimensional space $E_{p,q}$ with signature (p, q) and $p > 1, q > 1$. The norm of a vector is defined as, say

$$\mathbf{x}^2 \equiv x_1^2 + \ldots + x_p^2 - x_{p+1}^2 - \ldots - x_{p+q}^2, \tag{5.64}$$

and the scalar product as

$$\mathbf{x} \cdot \mathbf{y} \equiv x_1 y_1 + \ldots + x_p y_p - x_{p+1} y_{p+1} - \ldots - x_{p+q} y_{p+q}. \tag{5.65}$$

The one-vector orbits (for non-lightlike vectors) are

$$\mathcal{O}_{p=1,\text{timelike}} = \frac{SO\,(p,q)}{SO\,(p-1,q)} \quad \text{if } \mathbf{x}^2 > 0; \tag{5.66}$$

$$\mathcal{O}_{p=1,\text{spacelike}} = \frac{SO\,(p,q)}{SO\,(p,q-1)} \quad \text{if } \mathbf{x}^2 < 0. \tag{5.67}$$

It is intuitively clear that the two-vector orbits do depend on the nature of the vectors themselves. Let us start and consider two timelike vectors ($\mathbf{x}^2 > 0$ and $\mathbf{y}^2 > 0$), whose one-center orbits are separately given by $\mathcal{O}_{p=1,\text{timelike}}$. It is straightforward to show that the two-center orbits supporting this configuration are

$$\frac{SO\,(p,q)}{SO\,(p-2,q)} \quad \text{if } \mathbf{x}^2 \mathbf{y}^2 > (\mathbf{x} \cdot \mathbf{y})^2; \tag{5.68}$$

$$\frac{SO\,(p,q)}{SO\,(p-1,q-1)} \quad \text{if } \mathbf{x}^2 \mathbf{y}^2 < (\mathbf{x} \cdot \mathbf{y})^2. \tag{5.69}$$

If both vectors are spacelike ($\mathbf{x}^2 < 0$ and $\mathbf{y}^2 < 0$), the two-center orbits read

$$\frac{SO\,(p,q)}{SO\,(p,q-2)} \text{ if } \mathbf{x}^2\mathbf{y}^2 > (\mathbf{x}\cdot\mathbf{y})^2\,; \tag{5.70}$$

$$\frac{SO\,(p,q)}{SO\,(p-1,q-1)} \text{ if } \mathbf{x}^2\mathbf{y}^2 < (\mathbf{x}\cdot\mathbf{y})^2\,. \tag{5.71}$$

Finally, if one vector is timelike and the other one is spacelike (say, $-\mathbf{x}^2 > 0$ and $\mathbf{y}^2 < 0$), the two-center orbit is unique:

$$\frac{SO\,(p,q)}{SO\,(p-1,q-1)}\,, \tag{5.72}$$

because in this case $\mathbf{x}^2\mathbf{y}^2 < (\mathbf{x}\cdot\mathbf{y})^2$ always holds.

By introducing the $SL_h\,(2,R)\times SO\,(p,q)$ invariant polynomial (see [12,19] and the last Ref. of [22])

$$\mathbf{I}_4\,(\mathbf{x},\mathbf{y}) \equiv \mathbf{x}^2\mathbf{y}^2 - (\mathbf{x}\cdot\mathbf{y})^2\,, \tag{5.73}$$

all orbits (5.68)–(5.72) can actually be recognised to correspond to only three orbits (namely (5.68), (5.70), and (5.69)=(5.71)=(5.72)), respectively defined by the $[SL_h\,(2,R)\times SO\,(p,q)]$-invariant constraints: $\mathbf{I}_4 > 0$ (with $\mathbf{x}^2 > 0$ and $\mathbf{y}^2 > 0$); $\mathbf{I}_4 > 0$ (with $\mathbf{x}^2 < 0$ and $\mathbf{y}^2 < 0$); $\mathbf{I}_4 < 0$. Note that in the compact case (Euclidean signature: $q = 0$) $\mathbf{I}_4 > 0$ due to the Cauchy-Schwarz triangular inequality, and the two-vector orbit is unique: $\frac{SO(p)}{SO(p-2)}$. This is in analogy with the results (obtained in the complex field) discussed in Sect. 5.2.

5.4   Invariant Structures and the Role of the Horizontal Symmetry $SL_h\,(2,\mathbb{R})$

We now propose a candidate for a complete basis of G_4-invariant polynomials for the $p = 2$ case, highlighting the role of the horizontal symmetry group [26] in the classification of multi-center invariant structures.

Our treatment applies at least to the *irreducible* cubic geometries of *symmetric* scalar manifolds of $d = 4$ supergravity theories [15] (which, with the exception of the rank-1 t^3 model,[6] are the ones considered in the counting analysis of Sect. 5.2):

1. $\mathcal{N} = 2$ magical Maxwell-Einstein supergravities ($J_3^{\mathbb{A}}$, $\mathbb{A} = \mathbb{O}, \mathbb{H}, \mathbb{C}, \mathbb{R}$), with the case $J_3^{\mathbb{H}}$ encompassing also $\mathcal{N} = 6$ "pure" supergravity [1,3,24,31,37];
2. $\mathcal{N} = 5$ "pure" supergravity ($M_{1,2}\,(\mathbb{O})$);
3. $\mathcal{N} = 8$ "pure" supergravity ($J_3^{\mathbb{O}_s}$).

---

[6]As mentioned above, the irreducible rank-1 cubic case (the so-called $\mathcal{N} = 2$, $d = 4$ t^3 model, associated to the trivial degree-1 Jordan algebra $\mathbb{R}$) has been treated in [26].

The simplest invariant structures of a simple Lie group G (such as the U-duality group G_4 of an *irreducible* symmetric model) are the Killing-Cartan metric $g_{\alpha\beta}$, the structure constants $f_{\alpha\beta\gamma}$ and the symplectic metric $\mathbb{C}_{MN}$ (the Greek indices are in the adjoint representation of G_4, $\mathbf{Adj}\,(G_4)$, while the capital indices are in $\mathbf{Sympl}\,(G_4)$). It is well known that the entries of the generators in $\mathbf{Sympl}\,(G_4)$

$$t_{\alpha|MN} \equiv t_{\alpha|M}{}^P \mathbb{C}_{PN} = t_{\alpha|(MN)} \tag{5.74}$$

are invariant structures, symmetric in the symplectic indices (for the notation, see [14]).

In particular, one can construct the so-called K-tensor[7] [36]

$$\mathbb{K}_{MNPQ} \equiv -\frac{1}{3\tau} t_{(MN}^{\alpha} t_{\alpha|PQ)} = -\frac{1}{3\tau}\left(t_{MN}^{\alpha} t_{\alpha|PQ} - \tau\,\mathbb{C}_{M(P}\mathbb{C}_{Q)N}\right) = \mathbb{K}_{(MNPQ)}, \tag{5.75}$$

where τ is a G_4-dependent constant defined as

$$\tau \equiv \frac{2d}{f(f+1)}, \tag{5.76}$$

with $d \equiv \dim_{\mathbb{R}}\mathbf{Adj}\,(G_4)$ and $f \equiv \dim_{\mathbb{R}}(\mathbf{Sympl}\,(G_4))$. From its definition (5.75), the K-tensor is a completely symmetric rank-4 G_4-invariant tensor of $\mathbf{Sympl}\,(G_4)$.

In the presence of a single-centered black-hole background ($p = 1$) associated to a dyonic black-hole charge vector $\mathcal{Q}^M$ in $\mathbf{Sympl}\,(G_4)$, the unique independent G_4-invariant polynomial reads [36]

$$\mathfrak{I}_4\left(\mathcal{Q}^4\right) \equiv \mathbb{K}_{MNPQ}\mathcal{Q}^M\mathcal{Q}^N\mathcal{Q}^P\mathcal{Q}^Q = -\frac{1}{3\tau}t_{MN}^{\alpha}t_{\alpha|PQ}\mathcal{Q}^M\mathcal{Q}^N\mathcal{Q}^P\mathcal{Q}^Q. \tag{5.77}$$

On the other hand, in the presence of a multi-centered black-hole solution ($p \geqslant 2$), the horizontal symmetry $SL_h\,(p, \mathbb{R})$ [26] plays a crucial role in organizing the various G_4-covariant and G_4-invariant structures.

In the following we will consider the two-centered case ($p = 2$), the index $a = 1, 2$ spanning the fundamental representation (spin $s = 1/2$) $\mathbf{2}$ of the horizontal symmetry $SL_h\,(2, \mathbb{R})$.

By using the symplectic representation (5.74) of the generators of G_4, one can introduce the tensor (homogeneous quadratic in charges)

$$T_{\alpha|ab} \equiv t_{\alpha|MN}\mathcal{Q}_a^M\mathcal{Q}_b^N = T_{\alpha|(ab)} = \begin{pmatrix} T_{\alpha|11} & T_{\alpha|12} \\ T_{\alpha|12} & T_{\alpha|22} \end{pmatrix}, \tag{5.78}$$

---

[7]With respect to the treatment given in [36], we fix the overall normalization constant of the K-tensor to the value $\xi = -\frac{1}{3\tau} = -\frac{f(f+1)}{6d}$, as needed for consistency reasons.

lying in $(\mathbf{3}, \mathbf{Adj}\,(G_4))$ of $SL_h\,(2, \mathbb{R}) \times G_4$, where $\mathbf{3}$ is the rank-2 symmetric (spin $s = 1$) representation of $SL_h\,(2, \mathbb{R})$. In *irreducible* models, $T_{\alpha|ab}$ is the analogue of the so-called $\mathbb{T}$-tensor, introduced in [26] for *reducible* theories. Under the centers' exchange $1 \leftrightarrow 2$, $T_{\alpha|11} \leftrightarrow T_{\alpha|22}$, while $T_{\alpha|12}$ is invariant.

Interestingly, one can prove that the quantity

$$\mathbf{N} \equiv g^{\alpha\beta} \left(T_{\alpha|11} T_{\beta|22} - T_{\alpha|12} T_{\beta|12} \right) \tag{5.79}$$

is *not* independent from lower order invariants. Indeed, *at least* in the aforementioned irreducible cases, it holds that

$$t^{\alpha}_{M[N} t_{\alpha|P]Q} = \frac{\tau}{2} \left[\mathbb{C}_{M(P} \mathbb{C}_{Q)N} - \mathbb{C}_{M(N} \mathbb{C}_{Q)P} \right]. \tag{5.80}$$

Thus, from (5.78) and (5.79) it follows that

$$\begin{aligned} \mathbf{N} &= 2t^{\alpha}_{M[N} t_{\alpha|P]Q} \mathcal{Q}_1^M \mathcal{Q}_1^N \mathcal{Q}_2^P \mathcal{Q}_2^Q \\ &= -\frac{1}{3} \left[\mathbb{C}_{M(P} \mathbb{C}_{Q)N} - \mathbb{C}_{M(N} \mathbb{C}_{Q)P} \right] \mathcal{Q}_1^M \mathcal{Q}_1^N \mathcal{Q}_2^P \mathcal{Q}_2^Q \\ &= \frac{1}{2} \mathcal{W}^2, \end{aligned} \tag{5.81}$$

where

$$\mathcal{W} \equiv \langle \mathcal{Q}_1, \mathcal{Q}_2 \rangle \equiv \frac{1}{2} \mathbb{C}_{MN} \epsilon^{ab} \mathcal{Q}_a^M \mathcal{Q}_b^N \tag{5.82}$$

is the *symplectic product* of the charge vectors $\mathcal{Q}_1$ and $\mathcal{Q}_2$, which is a singlet $(\mathbf{1}, \mathbf{1})$ of $SL_h\,(2, \mathbb{R}) \times G_4$ (manifestly antisymmetric under $1 \leftrightarrow 2$).

An important difference between the *reducible* models (studied in [26]) and the *irreducible* treated in the present investigation is that, while the former generally have a non-vanishing horizontal invariant polynomial $\mathfrak{X}$, the latter have it vanishing identically. Indeed, the analogue of $\mathfrak{X}$ (defined by Eq. (4.13) of [26]) for irreducible models can be defined as

$$\mathfrak{X}_{irred} \equiv \mathbf{N} - \frac{1}{2} \mathcal{W}^2 = 0, \tag{5.83}$$

where result (5.81) was used in the last step. The t^3 model mentioned in the Introduction is a non-generic irreducible model (studied in Sect. 7 of [26]); in this case, the vanishing of $\mathfrak{X}$ is given by Eq. (7.16) of [26].

By using the K-tensor (5.75), one can also define the tensor (homogeneous cubic in charges)

$$\mathcal{Q}_{M|abc} \equiv \mathbb{K}_{MNPQ} \mathcal{Q}_a^N \mathcal{Q}_b^P \mathcal{Q}_c^Q = \mathcal{Q}_{M|(abc)}, \tag{5.84}$$

lying in $(\mathbf{4}, \mathbf{Sympl}\,(G_4))$ of $SL_h\,(2, \mathbb{R}) \times G_4$, where $\mathbf{4}$ is the rank-3 symmetric representation (spin $s = 3/2$) of $SL_h\,(2, \mathbb{R})$. Under $1 \leftrightarrow 2$, we have $\mathcal{Q}_{M|111} \leftrightarrow \mathcal{Q}_{M|222}$ and $\mathcal{Q}_{M|112} \leftrightarrow \mathcal{Q}_{M|122}$.

By further contracting with a two-centered charge vector, one can introduce the tensor (homogeneous quartic in charges)

$$\mathbf{I}_{abcd} \equiv \mathbb{K}_{MNPQ}\mathcal{Q}_a^M \mathcal{Q}_b^N \mathcal{Q}_c^P \mathcal{Q}_d^Q = \mathbf{I}_{(abcd)}, \tag{5.85}$$

lying in $(\mathbf{5}, \mathbf{1})$ of $SL_h\,(2, \mathbb{R}) \times G_4$, where $\mathbf{5}$ is the rank-4 symmetric representation (spin $s = 2$) of $SL_h\,(2, \mathbb{R})$. Under $1 \leftrightarrow 2$, $\mathbf{I}_{1111} \leftrightarrow \mathbf{I}_{2222}$, $\mathbf{I}_{1112} \leftrightarrow \mathbf{I}_{1222}$, while $\mathbf{I}_{1122}$ is invariant.

Trivially, $\mathcal{Q}_{M|abc}$ and $\mathbf{I}_{(abcd)}$ are related by[8]

$$\mathbf{I}_{abcd} = \mathcal{Q}_{M|abc}\mathcal{Q}_d^M = \mathbb{C}^{MN}\mathcal{Q}_{M|abc}\mathcal{Q}_{N|d} = \langle \mathcal{Q}_{abc}, \mathcal{Q}_d \rangle; \tag{5.86}$$

$$\mathcal{Q}_{M|abc} = \frac{1}{4}\frac{\partial \mathbf{I}_{abcd}}{\partial \mathcal{Q}_d^M}. \tag{5.87}$$

Note that only the completely symmetric part $\mathcal{Q}_{M|(abc}\mathcal{Q}_{d)}^M$ survives the contraction in (5.86), because $\mathcal{Q}_{M|abc}\mathcal{Q}_d^M \epsilon^{cd} = 0$ from the symmetry of the K-tensor (5.75) and the definition (5.84) of $\mathcal{Q}_{M|abc}$ itself.

In order to generate G_4-invariant polynomials, one can:

1. Multiply and contract on $\mathbf{Adj}\,(G_4)$ the three components of the quadratic tensor $T_{\alpha|ab}$ defined by (5.78), *or*
2. Contract all four components of $\mathcal{Q}_{M|abc}$ defined by (5.84) with three 2-center charge vectors, in all possible ways, *or*
3. Contract all five components of $\mathbf{I}_{abcd}$ defined by (5.85) with four 2-center charge vectors, in all possible ways.

By virtue of the various relations considered above, these three approaches give equivalent results, which we now specify for the sake of clarity:

$$\mathbf{I}_{+2}\left(\mathcal{Q}_1^4\right) \equiv \mathcal{I}_4\left(\mathcal{Q}_1^4\right) \equiv \mathbf{I}_{1111} = \left(\widetilde{\mathcal{Q}}_{111}, \mathcal{Q}_1\right) = \mathbb{K}_{MNPQ}\mathcal{Q}_1^M \mathcal{Q}_1^N \mathcal{Q}_1^P \mathcal{Q}_1^Q$$

$$= -\frac{1}{3\tau}T_{11}^\alpha T_{\alpha|11}; \tag{5.88}$$

$$\mathbf{I}_{+1}\left(\mathcal{Q}_1^3 \mathcal{Q}_2\right) \equiv \mathbf{I}_{1112} = \left(\widetilde{\mathcal{Q}}_{111}, \mathcal{Q}_2\right) = \left(\widetilde{\mathcal{Q}}_{112}, \mathcal{Q}_1\right) = \mathbb{K}_{MNPQ}\mathcal{Q}_1^M \mathcal{Q}_1^N \mathcal{Q}_1^P \mathcal{Q}_2^Q$$

$$= -\frac{1}{3\tau}T_{11}^\alpha T_{12|\alpha}; \tag{5.89}$$

---

[8]We remark that relation (5.87) characterizes $\mathcal{Q}_{abc}$ as the two-center generalisation of the so-called *Freudenthal dual* of the dyonic charge vector $\mathcal{Q}^M$, introduced (with a different normalisation) in [8]. Thus, $\mathcal{Q}_{abc}$ can be regarded as the (polynomial) two-*center Freudenthal dual* of the dyonic charge vector $\mathcal{Q}_d$. Furthermore, Eqs. (5.82), (5.86) and (5.97) imply that, under the formal interchange $\mathcal{Q}_a^M \leftrightarrow \mathbb{C}^{MN}\mathcal{Q}_{N|abc}$, $\mathbf{I}_{abcd}$ is invariant and $\mathcal{W} \leftrightarrow \mathbf{I}_6$.

$$\mathbf{I}_0\left(\mathfrak{Q}_1^2\mathfrak{Q}_2^2\right) \equiv \mathbf{I}_{1122} = \left(\widetilde{\mathfrak{Q}}_{112}, \mathfrak{Q}_2\right) = \left(\widetilde{\mathfrak{Q}}_{122}, \mathfrak{Q}_1\right) = \mathbb{K}_{MNPQ}\mathfrak{Q}_1^M\mathfrak{Q}_1^N\mathfrak{Q}_2^P\mathfrak{Q}_2^Q$$

$$= -\frac{1}{9\tau}\left(T_{11}^\alpha T_{22|\alpha} + 2T_{12}^\alpha T_{12|\alpha}\right) - \frac{1}{3\tau}\left(T_{11}^\alpha T_{22|\alpha} + \tau W^2\right); \quad (5.90)$$

$$\mathbf{I}_{-1}\left(\mathfrak{Q}_1\mathfrak{Q}_2^3\right) \equiv \mathbf{I}_{1222} = \left(\widetilde{\mathfrak{Q}}_{122}, \mathfrak{Q}_2\right) = \left(\widetilde{\mathfrak{Q}}_{222}, \mathfrak{Q}_1\right) = \mathbb{K}_{MNPQ}\mathfrak{Q}_1^M\mathfrak{Q}_2^N\mathfrak{Q}_2^P\mathfrak{Q}_2^Q$$

$$= -\frac{1}{3\tau}T_{22}^\alpha T_{12|\alpha}; \quad (5.91)$$

$$\mathbf{I}_{-2}\left(\mathfrak{Q}_2^4\right) \equiv \mathfrak{I}_4\left(\mathfrak{Q}_2^4\right) \equiv \mathbf{I}_{2222} = \left(\widetilde{\mathfrak{Q}}_{222}, \mathfrak{Q}_2\right) = \mathbb{K}_{MNPQ}\mathfrak{Q}_2^M\mathfrak{Q}_2^N\mathfrak{Q}_2^P\mathfrak{Q}_2^Q$$

$$= -\frac{1}{3\tau}T_{22}^\alpha T_{22|\alpha}. \quad (5.92)$$

The subscripts in the G_4-invariant polynomials $\mathbf{I}_{+2}$, $\mathbf{I}_{+1}$, $\mathbf{I}_0$, $\mathbf{I}_{-1}$ and $\mathbf{I}_{-2}$ defined by (5.88)–(5.92) denote the polarization with respect to the horizontal symmetry $SL_h(2, \mathbb{R})$, inherited from the components of $\mathbf{I}_{abcd}$ (5.85); indeed, the five G_4-invariant polynomials (5.88)–(5.92) sit in the rank-4 symmetric representation (spin $s = 2$) **5** of $SL_h(2, \mathbb{R})$ itself [26].

In order to proceed further, it is worth mentioning the decomposition [36]

$$t_{\alpha|M}^{N}t_{\beta|NQ} = -t_{\alpha|MP}t_{\beta|NQ}\mathbb{C}^{PN} = \frac{1}{2n}g_{\alpha\beta}\mathbb{C}_{MQ} + \frac{1}{2}f_{\alpha\beta}^{\gamma}t_{\gamma|MQ} + S_{(\alpha\beta)[MQ]}, \quad (5.93)$$

where

$$S_{\alpha\beta|MN} = S_{(\alpha\beta)|[MN]} \quad (5.94)$$

denotes an invariant primitive tensor of G_4. From (5.93), the following identity for the K-tensor can be derived [36] (recall Footnote 7):

$$\mathbb{K}_{MNPQ}\mathbb{K}_{RSTU}\mathbb{C}^{QR} = -\frac{(f+1)}{6d}\mathbb{K}_{(MN|(ST}\mathbb{C}_{U)|P)} +$$

$$+\frac{(f+1)}{18d}\mathbb{C}_{(M|(S|}\mathbb{C}_{|N||T|}\mathbb{C}_{|P)|U)} + \frac{f^2(f+1)^2}{72d^2}f_{\alpha\beta\gamma}t^\alpha_{(MN}t^\beta_{P)(S}t^\gamma_{TU)} +$$

$$-\frac{f^2(f+1)^2}{36d^2}t^\alpha_{(MN}S_{\alpha\beta|P)(S}t^\beta_{TU)}, \quad (5.95)$$

where

$$S_{\alpha\beta|12} \equiv S_{\alpha\beta|MN}\mathfrak{Q}_1^M\mathfrak{Q}_2^N = S_{\alpha\beta|MN}\mathfrak{Q}_1^{[M}\mathfrak{Q}_2^{N]} = -S_{\alpha\beta|21}. \quad (5.96)$$

A G_4-invariant polynomial homogeneous sextic in charges can then be defined as follows:

$$\mathbf{I}_6\left(\mathcal{Q}_1^3\mathcal{Q}_2^3\right) \equiv \frac{1}{8}\left\langle \mathcal{Q}_{abc}, \mathcal{Q}_{def}\right\rangle \epsilon^{ad}\,\epsilon^{be}\,\epsilon^{cf} = \frac{1}{8}\mathbb{C}^{MN}\mathcal{Q}_{M|abc}\mathcal{Q}_{N|def}\,\epsilon^{ad}\,\epsilon^{be}\,\epsilon^{cf}$$

$$= \frac{1}{4}\left\langle \mathcal{Q}_{111}, \mathcal{Q}_{222}\right\rangle + \frac{3}{4}\left\langle \mathcal{Q}_{122}, \mathcal{Q}_{112}\right\rangle$$

$$= \frac{1}{4}\mathbb{K}_{MNPQ}\mathbb{K}_{RSTU}\mathbb{C}^{QR}\Big(\mathcal{Q}_1^M\mathcal{Q}_1^N\mathcal{Q}_1^P\mathcal{Q}_2^S\mathcal{Q}_2^T\mathcal{Q}_2^U +$$

$$+3\mathcal{Q}_1^M\mathcal{Q}_2^N\mathcal{Q}_2^P\mathcal{Q}_1^S\mathcal{Q}_1^T\mathcal{Q}_2^U\Big)$$

$$= \frac{(f+1)}{36d}\mathcal{W}^3 + \frac{f^2\,(f+1)^2}{144d^2}f_{\alpha\beta\gamma}\,T_{11}^\alpha T_{12}^\beta T_{22}^\gamma +$$

$$+\frac{f^2\,(f+1)^2}{108d^2}\left(T_{12}^\alpha T_{12}^\beta - T_{11}^\alpha T_{22}^\beta\right)S_{\alpha\beta|12}. \tag{5.97}$$

Note that $\mathbf{I}_6$ is manifestly antisymmetric under $1 \leftrightarrow 2$. The first line of (5.97) is manifestly $[SL_h(2,\mathbb{R})\times G_4]$-invariant, the second and third lines provide explicit expressions, and in the fourth line the identity (5.95) was exploited.

The symplectic product $\mathcal{W}$ (defined in (5.82)) is different from zero when the two charge vectors $\mathcal{Q}_1^M$ and $\mathcal{Q}_2^M$ are *mutually non-local*. The concept of *mutual non-locality* is very important in the treatment of marginal stability in multi-center black holes (see e.g. [5, 10, 16–18, 28, 29]).

The above treatment suggests that a candidate for a complete basis of G_4-invariant polynomials in the irreducible cases under consideration is given by the seven polynomials:

$$\left(\mathcal{W},\ \mathbf{I}_{+2},\ \mathbf{I}_{+1},\ \mathbf{I}_0,\ \mathbf{I}_{-1},\ \mathbf{I}_{-2},\ \mathbf{I}_6\right), \tag{5.98}$$

respectively defined by (5.82), (5.88)–(5.92) and (5.97).

Let us now construct the corresponding candidate for a complete basis of $[SL_h(2,\mathbb{R})\times G_4]$-invariant polynomials in the irreducible cases under consideration. The spin $s = 2$ representation $\mathbf{5}$ of $SL_h(2,\mathbb{R})$, whose components are the G_4-invariant polynomials $\mathbf{I}_{+2}, \mathbf{I}_{+1}, \mathbf{I}_0, \mathbf{I}_{-1}$ and $\mathbf{I}_{-2}$ (defined by (5.88)–(5.92)), can be re-arranged as a 3×3 symmetric traceless matrix $\mathbf{I}$ [26]. The only independent $SL_h(2,\mathbb{R})$-singlets which can be built out of such a 3×3 symmetric matrix $\mathbf{I}$, due to its tracelessness are [26]:

$$\mathrm{Tr}\left(\mathbf{I}^2\right) = \mathbf{I}_{+2}\mathbf{I}_{-2} + 3\mathbf{I}_0^2 - 4\mathbf{I}_{+1}\mathbf{I}_{-1}; \tag{5.99}$$

$$\mathrm{Tr}\left(\mathbf{I}^3\right) = \mathbf{I}_0^3 + \mathbf{I}_{+2}\mathbf{I}_{-1}^2 + \mathbf{I}_{-2}\mathbf{I}_{+1}^2 - \mathbf{I}_{+2}\mathbf{I}_{-2}\mathbf{I}_0 - 2\mathbf{I}_{+1}\mathbf{I}_0\mathbf{I}_{-1}. \tag{5.100}$$

which are homogeneous of order 8 and 12 in the charges respectively. Note that $\mathrm{Tr}(\mathbf{I}^2)$ and $\mathrm{Tr}(\mathbf{I}^3)$ are both invariant under $1 \leftrightarrow 2$.

We conclude that a possible candidate for a complete basis of $[SL_h(2,\mathbb{R})\times G_4]$-invariant polynomials in the irreducible cases under consideration is then given by the four polynomials

$$\left(\mathcal{W}, \ \mathbf{I}_6, \ \mathrm{Tr}\left(\mathbf{I}^2\right), \ \mathrm{Tr}\left(\mathbf{I}^3\right)\right), \tag{5.101}$$

Note that, as shown by Kaç [33], the four invariants in Eq. (5.101) are a complete basis of finitely generated invariant polynomials. In particular, $\mathbf{I}_6$ is unique and this implies that there is one linear relation among the three structures which appear in Eq. (5.97).

It is worth pointing out that the analysis of Sect. 5.2 and of the present section can be easily generalized to $p \geq 3$ centers. The two-centered representation of spin $s = J/2$ of $SL_h (2, \mathbb{R})$ is then replaced by the completely symmetric rank-J tensor representation $\mathcal{R}_J$ of $SL_h (p, \mathbb{R})$ ($J = 1, 2, 3, 4$ are the values relevant for the above analysis). On the other hand, $\mathcal{W}$ and $\mathbf{I}_6$ generally sit in the $\left(\widetilde{\mathcal{R}}_2, \mathbf{1}\right)$ representation of $SL_h (p, \mathbb{R}) \times G_4$, where $\widetilde{\mathcal{R}}_2$ is the rank-2 antisymmetric representation of $SL_h (p, \mathbb{R})$ (which, in the case $p = 2$, becomes a singlet). However, due to the tree structure of the split flow in multi-center supergravity solutions [5, 16–18], to consider only the case $p = 2$ does not imply any loss in generality as far as marginal stability issues are concerned.

Acknowledgements The present contribution is based on [4] made in collaboration with Alessio Marrani and Mario Trigiante. The work of S. F. is supported by the ERC Advanced Grant no. 226455, *"Supersymmetry, Quantum Gravity and Gauge Fields"* (*SUPERFIELDS*). The work of L.A. and R.D'A. is supported in part by the MIUR-PRIN contract 2009-KHZKRX.

References

1. L. Andrianopoli, R. D'Auria, S. Ferrara, U-invariants, black hole entropy and fixed scalars. Phys. Lett. **B403**, 12 (1997). hep-th/9703156
2. L. Andrianopoli, R. D'Auria, S. Ferrara, M. Trigiante, Extremal black holes in supergravity. Lect. Notes Phys. **737**, 661 (2008). hep-th/0611345
3. L. Andrianopoli, R. D'Auria, S. Ferrara, P.A. Grassi, M. Trigiante, Exceptional $\mathcal{N} = 6$ and $\mathcal{N} = 2 \, AdS_4$ supergravity, and zero-center modules. J. High Energy Phys. **0904**, 074 (2009). arXiv:0810.1214 (hep-th)
4. L. Andrianopoli, R. D'Auria, S. Ferrara, A. Marrani, M. Trigiante, Two-centered magical charge orbits. J. High Energy Phys. **1104**, 041 (2011). (arXiv:1101.3496 (hep-th))
5. B. Bates, F. Denef, *Exact Solutions for Supersymmetric Stationary Black Hole Composites.* J. High Energy Phys. **1111**, 127 (2011). hep-th/0304094
6. J.C. Baez, The octonions. Bull. Am. Math. Soc. **39**, 145 (2001). math/0105155
7. S. Bellucci, S. Ferrara, M. Günaydin, A. Marrani, Charge orbits of symmetric special geometries and attractors. Int. J. Mod. Phys. **A21**, 5043 (2006). hep-th/0606209
8. L. Borsten, D. Dahanayake, M.J. Duff, W. Rubens, Black holes admitting a freudenthal dual. Phys. Rev. **D80**, 026003 (2009). arXiv:0903.5517 (hep-th)
9. G. Bossard, *1/8 BPS Black Hole Composites.* arXiv:1001.3157 (hep-th)
10. A. Castro, J. Simon, Deconstructing the *D0-D6* system. J. High Energy Phys. **0905**, 078 (2009). arXiv:0903.5523 (hep-th)
11. E. Cremmer, B. Julia, The $\mathcal{N} = 8$ supergravity theory. 1. The lagrangian. Phys. Lett. **B80**, 48 (1978); E. Cremmer, B. Julia, The $SO(8)$ supergravity. Nucl. Phys. **B159**, 141 (1979)

12. M. Cvetic, D. Youm, Dyonic BPS saturated black holes of heterotic string on a six torus. Phys. Rev. **D53**, 584 (1996), hep-th/9507090; M. Cvetic, A.A. Tseytlin, General class of BPS saturated dyonic black holes as exact superstring solutions. Phys. Lett. **B366**, 95 (1996). hep-th/9510097; M. Cvetic, A.A. Tseytlin, Solitonic strings and BPS saturated dyonic black holes. Phys. Rev. **D53**, 5619 (1996); Erratum-ibid. **D55**, 3907 (1997), hep-th/9512031

13. J.R. David, On walls of marginal stability in $\mathcal{N} = 2$ string theories. J. High Energy Phys. **0908**, 054 (2009). arXiv:0905.4115 (hep-th)

14. B. de Wit, H. Samtleben, M. Trigiante, On lagrangians and gaugings of maximal supergravities. Nucl. Phys. **B655**, 93 (2003). hep-th/0212239

15. B. de Wit, F. Vanderseypen, A. Van Proeyen, Symmetry structures of special geometries. Nucl. Phys. **B400**, 463 (1993). hep-th/9210068

16. F. Denef, Supergravity flows and D-brane stability. J. High Energy Phys. **0008**, 050 (2000). hep-th/0005049

17. F. Denef, G.W. Moore, *Split States, Entropy Enigmas, Holes and Halos*. hep-th/0702146; J. High Energy Phys. **1111**, 129 (2011). F. Denef, D. Gaiotto, A. Strominger, D. Van den Bleeken, X. Yin, *Black Hole Deconstruction*. hep-th/0703252; J. High Energy Phys. **1203**, 071 (2012). F. Denef, G.W. Moore, How many black holes fit on the head of a pin? Gen. Relat. Gravit. **39**, 1539 (2007). arXiv:0705.2564 (hep-th)

18. F. Denef, B.R. Greene, M. Raugas, Split attractor flows and the spectrum of BPS D-branes on the quintic. J. High Energy Phys. **0105**, 012 (2001). hep-th/0101135

19. M.J. Duff, J.T. Liu, J. Rahmfeld, Four-dimensional string/string/string triality. Nucl. Phys. **B459**, 125 (1996). hep-th/9508094

20. S. Ferrara, M. Günaydin, Orbits of exceptional groups, duality and BPS states in string theory. Int. J. Mod. Phys. **A13**, 2075 (1998). hep-th/9708025

21. S. Ferrara, A. Marrani, Matrix norms, BPS bounds and marginal stability in $\mathcal{N} = 8$ supergravity. J. High Energy Phys. (2010, in press). arXiv:1009.3251 (hep-th)

22. S. Ferrara, R. Kallosh, A. Strominger, $\mathcal{N} = 2$ extremal black holes. Phys. Rev. **D52**, 5412 (1995). hep-th/9508072; A. Strominger, Macroscopic entropy of $\mathcal{N} = 2$ extremal black holes. Phys. Lett. **B383**, 39 (1996). hep-th/9602111; S. Ferrara, R. Kallosh, Supersymmetry and attractors. Phys. Rev. **D54**, 1514 (1996). hep-th/9602136; S. Ferrara, R. Kallosh, Universality of supersymmetric attractors. Phys. Rev. **D54**, 1525 (1996). hep-th/9603090

23. S. Ferrara, G.W. Gibbons, R. Kallosh, Black holes and critical points in moduli space. Nucl. Phys. **B500**, 75 (1997). hep-th/9702103

24. S. Ferrara, E.G. Gimon, R. Kallosh, Magic supergravities, $\mathcal{N} = 8$ and black hole composites. Phys. Rev. **D74**, 125018 (2006). hep-th/0606211

25. S. Ferrara, A. Marrani, E. Orazi, *Split Attractor Flow in $\mathcal{N} = 2$ Minimally Coupled Supergravity*. arXiv:1010.2280 (hep-th), Nucl. Phys. **B846**, 512–541 (2011)

26. S. Ferrara, A. Marrani, E. Orazi, R. Stora, A. Yeranyan, *Two-Center Black Holes Duality-Invariants for stu Model and its lower-rank Descendants*. arXiv:1011.5864 (hep-th), J. Math. Phys. **062302**, 52 (2011)

27. S. Ferrara, A. Marrani, A. Yeranyan, On *Invariant Structures of Black Hole Charges*. arXiv:1110.4004 (hep-th), J. High Energy Phys. **1202**, 071 (2012).

28. D. Gaiotto, W.W. Li, M. Padi, Non-supersymmetric attractor flow in symmetric spaces. J. High Energy Phys. **0712**, 093 (2007). arXiv:0710.1638 (hep-th)

29. E.G. Gimon, F. Larsen, J. Simon, Constituent model of extremal non-BPS black holes. J. High Energy Phys. **0907**, 052 (2009). arXiv:0903.0719 (hep-th)

30. M. Günaydin, *Lectures on Spectrum Generating Symmetries and U-Duality in Supergravity, Extremal Black Holes, Quantum Attractors and Harmonic Superspace*. arXiv:0908.0374 (hep-th)

31. M. Günaydin, G. Sierra, P.K. Townsend, Exceptional supergravity theories and the magic square. Phys. Lett. **B133**, 72 (1983); M. Günaydin, G. Sierra, P.K. Townsend, The geometry of $\mathcal{N} = 2$ Maxwell-Einstein supergravity and Jordan Algebras. Nucl. Phys. **B242**, 244 (1984)

32. C. Hull, P.K. Townsend, Unity of superstring dualities. Nucl. Phys. **B438**, 109 (1995). hep-th/9410167
33. V.G. Kaç, Some remarks on nilpotent orbits. J. Algebr. **64**, 190–213 (1980)
34. P. Levay, Two-center black holes, qubits and elliptic curves. Phys. Rev. D **84**, 025023 (2011). arXiv:1104.0144 (hep-th)
35. J.F. Luciani, Coupling of O(2) supergravity with several vector multiplets. Nucl. Phys. **B132**, 325 (1978)
36. A. Marrani, E. Orazi, F. Riccioni, Exceptional reductions. arXiv:1012.5797v1 (hep-th), J. Phys. A. **A44**, 155207 (2011)
37. D. Roest, H. Samtleben, Twin supergravities. Class. Quantum Gravity **26**, 155001 (2009). arXiv:0904.1344(hep-th)
38. R. Slansky, Group theory for unified model building. Phys. Rep. **79**, 1 (1981)

Printed by Publishers' Graphics LLC